AF522049

Washington 101

Washington 101

An Introduction to the Nation's Capital

Matthew N. Green, Julie Yarwood,
Laura Daughtery, and Maria Mazzenga

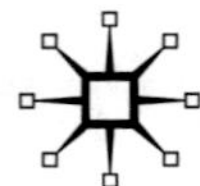

First published in 2014 by PALGRAVE MACMILLAN® in the United States—a division of St. Martin's Press LLC, 175 Fifth Avenue, New York, NY 10010.

Where this book is distributed in the UK, Europe and the rest of the world, this is by Palgrave Macmillan, a division of Macmillan Publishers Limited, registered in England, company number 785998, of Houndmills, Basingstoke, Hampshire RG21 6XS.

Palgrave Macmillan is the global academic imprint of the above companies and has companies and representatives throughout the world.

ISBN: 978-1-137-43338-1

Library of Congress Cataloging-in-Publication Data

Green, Matthew N., 1970–
Washington 101 : an introduction to the nation's capital / by Matthew N. Green, Julie Yarwood, Laura Daughtery, Maria Mazzenga.
pages cm
Includes bibliographical references and index.
ISBN 978-1-137-43338-1 (hardback)
1. Washington (D.C.)—History. 2. Washington (D.C.)—Description and travel. 3. Washington (D.C.)—Buildings, structures, etc. I. Title.
F194.G74 2014
975.3—dc23

2013050574

A catalogue record of the book is available from the British Library.

Design by Amnet.

First edition: June 2014

10 9 8 7 6 5 4 3 2 1

Contents

Part III: Washington as Living City

List of Figures

List of Tables

Acknowledgments

There are many people to thank for helping us with this project. Larry Poos, dean of Arts and Sciences at Catholic University, provided institutional support and leadership to create the interdisciplinary course that inspired this study. Timothy Meagher headed the committee that developed the class and was one of its first instructors. Other CUA faculty helped create and/or teach the course, including Thomas Donahue, Taryn Okuma, and Leslie Tentler. Thanks also to the many guest speakers who visited the class and offered important insights and experiences that informed our research, and to a number of our students who provided helpful feedback on the class.

The book might never have come to fruition had it not been for a chance encounter with Jeremy Weinstein of Stanford University, who suggested that we put what we had learned about Washington, D.C., onto the written page. Some of our research was conducted with the aid of a grant from the Institute for Policy Research and Catholic Studies at Catholic University. Holly Green and Terrance Williams were among those who read and commented on drafts of various chapters. For offering ideas, suggestions, research material, or encouragement, we thank Carolina Burner, John DeCrosta, Heather Gonzalez, John Kromkowski, Emery Lee, Frances Lee, John Mann, Jenifer Nawrocki, Enrique Pumar, Charlotte Sage, and Carol Young. Additional research assistance was provided by Greg Collins. Our editor at Palgrave, Brian O'Connor, has been wonderfully supportive of the project and helped bring about its timely approval.

Finally, we want to thank our friends and family members for their support and patience throughout the process of researching and writing this book.

Introduction

We have a tendency to typecast our cities. New York is a center of finance and ethnic diversity. San Francisco is where liberal bohemians gather. Seattle has rain, coffee shops, and the Space Needle. Miami is the land of beachgoers and Cuban émigrés.

Washington, D.C., America's capital city, evokes a very different set of images. Bold neoclassical monuments. The White House. Free museums. Bureaucrats and power-seeking politicians. These impressions, like those of other cities, are often based in reality. Yet they also hide a great deal of complexity. For instance, although Washington does house the national government and lots of elected officials, far more jobs in the area are actually created by the private sector, not the federal government. While Washington's residents live in the capital of the world's most powerful democracy, they lack full democratic representation. Though the city features many of America's most famous white classical monuments, Washington is also the home of the black modernist Vietnam Veterans Memorial and scores of bronze man-on-horseback statues. Nor do many realize that the city has long been one of the most important centers of African American life, or that the region is among the top immigrant destinations in the United States. To put it in frequently used shorthand, most people know about "Washington," a place of landmarks and tourist sites, but know little—or assume the worst—about "D.C."[1]

The primary purpose of this book is to dispel the myths and stereotypes of Washington, D.C., and reveal the multifaceted nature of its architecture, politics, economics, history, and people. Hundreds of books, some quite outstanding, have been written about the city. But most fall into one of two categories: tourist guides written to help the casual visitor navigate Washington's landmarks or academic studies of important but specialized topics. What is missing is a contemporary work that adequately explains Washington's core features—what the city is truly *about*.

The absence of such a book became clear to us when we were invited to be part of a team to develop and teach a new undergraduate class on the city of

Washington at the Catholic University of America, located in northeast D.C. and where we work as faculty. After struggling to find a single text to assign students, we decided to write one ourselves, one that would synthesize what has already been written about Washington while adding some perspective of our own. In your hands is the result of that endeavor.

Three features of Washington—features which have, at least individually, been noted by others as well—make the city and region unique among American urban places.[2] First, the city is *largely—though not entirely—shaped by national politics.* National politics has motivated the construction of scores of neoclassical buildings and monuments and nationally themed museums in Washington. It fueled the rise of the city as a hub of the African American community. It permeates the capital's elite society, influences the decisions of local government, explains the city's unique lack of independence, contributes substantially to the regional economy, and makes D.C. a magnet for protestors of all stripes. Washington would not even exist but for the decision of the country's founders to create a place, largely from scratch, to house a new national government. Washington, like all "great capital cities," is where "a government displays its personality," in the words of historian Alan Lessoff. As another scholar wrote, "American democracy is an artifact, and Washington DC is its concrete if imperfect realization."[3]

There is more to Washington than national politics, however. The region is home to major corporations that rely indirectly, if at all, on the government to earn their revenue or justify their corporate mission. The city's geographic location near the American South has been as important as its political status in contributing to Washington's southern culture and the formation of a large and prominent black population.[4] Also, many communities within Washington have little to do with national politics. The recollection of a woman who grew up in the Capitol Hill neighborhood in the 1930s still rings true for many Washingtonians today: "We knew the Capitol was there, but we had no dealings with it from day to day. We had our own separate lives—our own stores, our own businesses."[5]

The city's second distinctive feature is that it both *garners unusually generous benefits and faces unusually stringent limits* as the national capital. Political scientist Margaret Farrar puts it best when she writes that the city both "enjoys" and "endures" its status.[6] Special opportunities and benefits include a large supply of secure, well-paying government jobs; federal dollars that support local companies, think-tanks, and businesses; funding for world-class museums, which in turn contribute to a lucrative tourism industry; and a national audience for groups seeking attention to their favored cause. Constraints and drawbacks include an overreliance on government largess, which can be dangerous during periods of budget cutbacks; constitutionally fixed city boundaries, which hindered Washington's ability to expand its size and thereby capture the wealth of its immediate suburbs; and a lack of political independence and full representation in Congress. The last is perhaps the most galling to Washington residents: in no other city in the country does the national government have so much say over local laws, spending, and revenue, making D.C. very much a dependent center of American politics.

The third element distinguishing Washington from other cities is its long-standing and prominent status as a *crossroads of the African American experience.* Located south of the Mason-Dixon Line, the District of Columbia was founded in slave country, and the cities within it (Washington, Georgetown, and Alexandria) developed important slave-trading commercial activity. But the area also attracted thousands of freedmen, and many local slaves were able to purchase or otherwise legally obtain their freedom. The African American population continued to grow during and after the Civil War, and again during the Great Migration of the early 1900s. Though subject to considerable racial discrimination and prejudice—prejudice so powerful that it helped convince Congress to abolish local government altogether rather than let blacks continue to vote—African Americans in Washington formed a community that became a national center of black culture, scholarship, and civil rights activism. By 1960 Washington was the first non-southern American city to boast a majority black population, and the black community continues to be a tremendous influence on the city's politics, economy, and society.

The book's ten chapters are grouped by theme. The first three chapters are organized under the theme of Washington as a "symbolic city" and examine those features of the city that have the most symbolic import in appearance and design: its architecture, monuments, and museums. The next set of chapters explores the "political city," including the operation of the national government within the District's borders, the city's attraction to protestors, the contribution of embassies and international organizations to city life, and Washington's local politics. The final set of chapters fall under the rubric of the "living city." They review features of the local economy; regional demographics, with particular attention to the city's immigrant population and African American community; and the development and identity of city neighborhoods and suburban communities around Washington, D.C.

We acknowledge that these three divisions, while useful as a means of organizing the book, are to some extent arbitrary and porous. Sometimes the same place, phenomenon, or event touches upon all three "kinds" of a city. For instance, when Washington adopted a largely "southernized social scene" and identity in the 1850s, that aspect of the "living city" had both political ramifications (with southerners of the social scene frequently in positions of political power) and symbolic ones (since a southern identity of the capital

Figure I.1 Ethiopian Protest in Downtown Washington, May 2012
Photo by Matthew Green

"implied that southern values were national values").[7] Or take a more recent event: a May 2012 downtown protest against the autocratic rule of Ethiopian Prime Minister Meles Zenawi. The event highlighted Washington as not only a political city, but also as a symbolic city—the protest was held, fittingly, at Freedom Plaza, a public space dedicated to the founding of the capital—and as a living city, given that the protestors were mostly Ethiopian Americans, one of the city's most significant immigrant populations (see figure I.1).

Of course, we cannot provide a complete summary of all things Washington in a single book. A city is much more than its symbolic, political, and living elements. There is, for example, the "environmental city": an urban place's natural topography, weather, wildlife, and resources, which can affect everything from how attractive it is to tourists to the local political agenda.[8] Space also precludes us from dedicating attention to specific policy issues, such as crime and education, and from covering all of the topics that fall under the "living city" theme in particular, such as the arts.[9] We must leave that to another volume.

An important objective of our course at Catholic University, and of this book, is to put the city of Washington into a larger scholarly context. We are not urban theorists; we make no effort to place the city in an existing "school" of urban studies (like the Chicago School or the New York School) or make it the basis for a new one.[10] But Washington's similarities to, and differences from, other American cities do illustrate important concepts from the social sciences and humanities, including political science, economics, sociology, architecture, and ethnography. Accordingly, we introduce such concepts from time to time as a means of further illuminating what Washington, D.C., is all about.

Two other approaches that we take should also be mentioned. While this is not a work of history, we recognize that history helps us understand why a city looks and works as it does today. "Cities encapsulate time," as the historian Kevin Starr has written.[11] Accordingly, each chapter includes, in varying detail, some historical background of its topic. Second, the reader will note the occasional reference to our university and the nearby neighborhood of Brookland. Though this is perhaps a consequence of parochialism, we also believe that Catholic University and Brookland have been important and largely unrecognized elements of Washington's politics, culture, and history. For instance, few realize that Catholic University was, for many years, the only non-black university in the city to admit African Americans, though it would succumb to pressure to exclude blacks in the 1910s and 1920s, a testament to the power of racism and segregation in the city at that time.[12]

Above all, we hope we have written a book that pleases as much as it instructs. Half a million people live within Washington, D.C.'s boundaries, and nearly six million live in the greater metropolitan area. Together, they make up one of the most diverse, lively, and complicated urban areas in America. We see in Washington a fascinating urban place with a compelling story, and we hope that, after reading this book, you will too.

PART I

Washington as Symbolic City

Introduction to Part I

Washington, D.C., is as much a symbol as it is a real city. Much of that symbolism derives from its status as a capital: even its very name is used as shorthand to describe the U.S. government (and how Americans view their government—usually negatively). But it is also a symbol for other reasons, including its distinctive appearance and its public places and spaces.

In the next three chapters, we consider Washington, D.C., as a symbolic city. We look first to the city's architecture, especially its neoclassical public buildings and its highly geometric baroque street design. Both styles lend themselves to various interpretations of their "true" meaning, especially because together they give Washington an appearance unique among American cities. But in fact, what people think they stand for—democracy, power, religious (or religious-like) faith—were far from the minds of those who first brought those styles to the capital.

In chapter 2 we examine the preponderance of monuments and memorials in Washington. Taken collectively, they suggest that the city represents the nation at large and that it serves as a place of national remembrance. Yet there is considerable diversity in the style and structure of the city's memorials, due not only to changing architectural tastes but to significant shifts in who Americans believe should be remembered and how. Even the same memorial can come to symbolize different values over time.

Chapter 3 explores another feature of Washington that distinguishes it from most other American cities: the presence of many, often world-class, museums. Like its monuments and memorials, Washington's museums encourage a perception of the city as a national urban place in which America's historic and cultural treasures are kept and displayed with pride. As we shall see, the multifaceted task of the city's museums—to remember, to educate, to attract visitors, to preserve national identity—has sometimes led to controversy and conflict.

CHAPTER 1

Rome on the Potomac: The Classical Architecture of Washington

First impressions of a city are shaped by architecture, and Washington, D.C., evokes a powerful impression indeed. The city's tallest structure is the Washington Monument, an Egyptian-style obelisk that looms far above all else. Bold columns and grand archways decorate white stone buildings throughout the capital. The city's grid street pattern is overlaid with diagonal avenues and traffic circles, which, while at times confusing to navigate, direct the viewer's eyes to many impressive buildings. It is no wonder that the city has been called "Rome on the Potomac."

Washington actually features a diversity of building styles, and not all of its streets are broad, straight avenues.[1] But two architectural features in particular distinguish the city's central core[2] from the urban centers of other American cities. First, the layout of central Washington is *baroque*: broad, diagonal avenues cut across a traditional street grid, with important structures, monuments, and squares located where multiple streets intersect (see textbox 1). Second, its public buildings, particularly those located on or near the National Mall, are mostly *neoclassical* in design, featuring columns, white or grey stone surfaces, and other elements from the great buildings of Ancient Greece and Rome (see textbox 2).

TEXTBOX 1: Features of the Baroque City Plan

The architectural scholar Spiro Kostof neatly summarized how Washington exemplifies several principles of Baroque planning, including the following:

1. a total, grand, spacious urban ensemble pinned on focal points distributed throughout the city

2. these focal points suitably plotted in relation to the drama of the topography, and linked with each other by swift, sweeping lines of communication
3. a concern with the landscaping of the major streets . . .
4. the creation of vistas
5. public spaces as setting for monuments
6. dramatic effects, as with waterfalls and the like
7. all of this superimposed on a closer-grained fabric for daily, local life.[3]

Some of these features, particularly the broad streets and creation of focal points, can be seen in L'Enfant's original map (see figure 1.1).

TEXTBOX 2: Features of Neoclassical Architecture

As its name suggests, Neoclassicism, also sometimes called Classic Revival, is a style modeled after the great buildings of Ancient Athens and Rome. There are, in fact, many specific styles that are called "neoclassical,"[4] but they all tend to share several features.

- Overall, neoclassical buildings are usually laid out *symmetrically* along one, and occasionally two, axes. They also make *use of careful proportions* in the relationships between height, depth, length, and other features.
- Neoclassical buildings often feature *uniform stone surfaces*, usually white or grey, in imitation of the appearance of contemporary ruins from the ancient world.
- "The heart of all classical architecture," writes the scholar Robert Adam, is the use of *columns of one or more major "orders."* The easiest way to distinguish the orders is by the capital (top) of the column; they range from the unadorned Doric to the curled Ionic and the highly ornate Corinthian. Other orders include the Tuscan and Composite.
- *Arches* and *domes* often appear in neoclassical buildings as well, particularly those seeking to replicate the style of Roman architecture.[5]

To many visitors, Washington's Old World look symbolizes power, fitting for a city founded as a political capital. In fact, both neoclassicism and the baroque style were primarily introduced for other reasons, not least for their intrinsic elegance and beauty and their popularity at the time the city was established. One of the great strengths of both architectural styles, however, is their *symbolic flexibility*, which allows successive generations to find new meanings—such as political authority—in Washington's physical appearance. That flexibility, along with political and artistic leadership by individuals and institutions at key moments in the city's history, encouraged others to adopt or maintain neoclassicism and a baroque layout as the capital grew.

In this chapter we discuss the introduction of both styles to Washington, due largely to the leadership and vision of two individuals: Pierre L'Enfant and Thomas Jefferson. We then explore the varying associations that people have with neoclassicism and baroque architecture, illustrating how those styles can take on multiple symbolic meanings. Finally, we briefly review the waxing and waning of public support for the baroque and neoclassical styles throughout the city's history, with particular attention to how the influential McMillan Plan of 1901 reinforced both at a critical historical moment.

The Creation of a Planned City

All cities are, to some extent, the result of human design and intent. But certain ones are the clear consequence of abstract maps imposed on a geographical space. Some plans impose order on existing disorder, as, for example, those of the nineteenth-century French planner Baron Haussmann, who replaced curving streets in Paris with straight roads and boulevards. Others create order afresh: Roman and other ancient civilizations laid out new cities in grids, for instance, as did settlers of early American towns.[6]

The city of Washington was planned because it was new, but creating a capital from scratch was not foreordained. Several existing towns and cities were contemplated as the seat of the federal government.[7] However, George Washington was one of many former colonists who was suspicious of placing the capital in an existing urban place; as he put it, "The tumultuous populace of large cities are ever to be dreaded." In 1783, a near rebellion of disgruntled former soldiers in Philadelphia, Congress's temporary meeting place and a leading contender for the country's capital, did little to counter this sentiment.[8]

Residents of Southern states wanted a more southerly location for the capital, so that "southern views on any issue—including slavery—would be more readily heard than northern ones." They won, and, in 1790, after years of debate, Congress authorized a square, ten-by-ten mile territory of fixed boundaries to be created on the Potomac River. The territory's exact location, determined by George Washington himself, had political, economic, and military advantages. It was midway between the North and South, along an important highway connecting both regions, bisected by a potential commercial waterway with two important riverside towns (Georgetown and Alexandria), and far enough from the ocean to be deemed safe from naval attack.[9]

For the city's designer, Washington selected Pierre L'Enfant, a former Revolutionary Army soldier and practiced architect. It was a fateful choice. A man of "grand visions," L'Enfant described the ubiquitous grid pattern of the typical city as "tiresome and insipid" and believed the new country deserved something more creative. His proposed city encompassed over 5,000 acres, as much territory as was then covered by Philadelphia, New York, and Boston put together. It embodied numerous features of baroque city design, including broad and straight streets, major squares and plazas, and avenues that connect important monuments or buildings (see figure 1.1). The intended effect was to join multiple distant points together and create impressive views of particular objects or scenes. Though L'Enfant insisted his plan was, as he put it, "wholly new," it had some obvious parallels with, and may have borrowed from, several European cities (both actual and proposed) and settlements in North America.[10]

The plan featured two main axes. A primary axis ran from Capitol Hill to a point just south of the president's residence, close to where the Washington

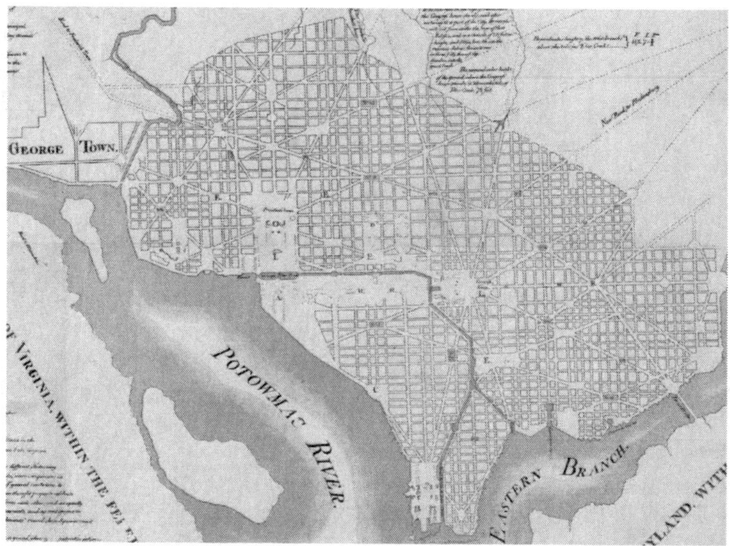

Figure 1.1 L'Enfant's Original Plan of Washington

Source: Geography and Map Division, Library of Congress

Monument stands today, and a promenade (now the National Mall) followed this line. The second axis was a major avenue connecting the Capitol to the home of the president (today's Pennsylvania Avenue). The result would be a "sense of order in, and of tactile command over, a large organism of space and solid," and by overlaying diagonal avenues on a traditional Cartesian grid, distant places would be joined to create what L'Enfant called "a reciprocity of sight."[11]

Explaining L'Enfant's Map

Why did L'Enfant design the city as he did? In the late eighteenth century the baroque style had become quite fashionable in Europe, but L'Enfant was driven by more than artistic trends. For one thing, he believed that a city should be *harmonious with existing topography*. He rejected a basic street grid for failing to take advantage of the area's hills, ridges, and waterways, and he dedicated its higher points to important government buildings like the Capitol. Certain city streets, such as Twelfth Street southwest and the southwest end of Maryland Avenue, would also feature views of the Potomac River.[12]

In addition, L'Enfant believed a city should *evoke a sense of grandeur*. The city of Washington, he wrote, "must leave to posterity a grand idea of the patriotic interest which promoted it," "engrave in every mind that sense of respect" owed to the capital of the new country, and be "of a magnitude so worthy to concern a grand empire" that it would make other nations jealous. Grandeur would come in part from the city's tremendous size, "proportioned to the greatness which . . . the Capital of a powerful Empire ought to manifest." But L'Enfant also included waterfalls, fountains, and a picturesque canal in his plan, and he expected landmarks to be placed at major intersections.[13]

Finally, L'Enfant felt that a city should *function well*, providing for the basic needs of its residents with room to grow economically and spatially. He proposed the construction of useful institutions like a national bank, church, theater, and markets, and provided space to add more such facilities plus "aggrandizement & embellishment" as might be needed in the future. L'Enfant expected the diagonal streets to encourage development at the city's distant edges, while canals and public squares along the Potomac River would ensure ready access to waterways, the primary avenues of commerce at the time. The architect also created fifteen squares, one for each state, intended to attract citizens from those states and, by exploiting interstate competition, spur growth around each.[14]

L'Enfant's motives mirrored the three purposes of architecture famously propounded by the Ancient Roman architect Vitruvius: durability (*firmitas*), usefulness (*utilitas*), and beauty (*venustas*).[15] The French architect believed his city had the necessary grandeur and ability to grow so as to endure; met the economic, spiritual, and recreational needs of its residents; and achieved a level of beauty through natural views and grand buildings. Other rumored purposes of his plan—to build a complex street pattern that would bewilder would-be invaders or to construct diagonal avenues to ease the deployment of troops against protestors—are urban legends.

As widely admired as it is, L'Enfant's plan has not been free from criticism. The intersections of diagonal avenues with the street grid can be confusing, even dangerous, and one architect complained that they "have their identity, beauty, and dignity sapped away by the constant intrusion, often at very acute angles, of the gridiron streets." Conventional development proved difficult on the odd-shaped plots created by these peculiar intersections, sometimes resulting in buildings that faced the street at odd angles. The great distance between Congress and the White House (as the president's home was informally called in the nineteenth century, and officially named by President Teddy Roosevelt in 1901) minimizes the strength of their axial connection. And the streets consume so much real estate that it was unlikely the city could have housed as many people as L'Enfant envisioned without the invention of taller buildings.[16]

Perhaps most damning is the complaint lodged by the urban critic Lewis Mumford that all baroque cities, including Washington, are hostile to everyday living. In a baroque plan, Mumford argued, each major avenue is so wide as to become a "barrier between its opposite sides" and invite vehicular rather than pedestrian traffic. By contrast, neighborhoods with shorter and narrower streets are attractive to live and shop in, Georgetown being a classic example. Defenders of L'Enfant's plan have countered that such criticism, even if valid, is more applicable to how the plan was implemented than to its original conception, which included more narrow, and maybe even some slightly crooked, avenues.[17]

The Introduction of Neoclassicism

L'Enfant did not serve as city planner long enough to suggest a dominant style for his city's main buildings. Onerous demands were continually placed on the French-born architect, and he in turn could be stubborn, unrealistic, and obsessed with the integrity of his map. After a series of missteps by L'Enfant—including dismantling the partially built home of a wealthy resident because it would jut onto a major street—President Washington relieved L'Enfant of his post in early 1792, barely one year after his appointment.[18]

Nonetheless, it seems likely that L'Enfant would have endorsed neoclassicism for the capital's buildings. In the late 1700s the style was admired in England and, as a consequence, increasingly popular in North America as well. Its appearance in Washington was due not only to British fashion, however, but to one man in particular: Thomas Jefferson. Jefferson, a skillful architect in his own right, helped popularize the idea of reviving older forms of architecture in America by designing highly praised neoclassical buildings. One of his most famous, Virginia's state capitol (1789), was the first to use the "pure classical temple form" in the United States. The building influenced Benjamin Latrobe, a major figure in early American architecture who became surveyor of Washington's public buildings in 1803. Latrobe added neoclassical elements to the White House and the Capitol building and introduced the idea of eclecticism—mixing different styles together—to American architecture.[19] In addition, Jefferson "hover[ed] constantly in the background of all government architectural activity" and helped select the final design of the Capitol after recommending that the structure should be, in his words, "predicated on antiquity." Jefferson was also responsible for Latrobe's appointment as Capitol architect in 1803.[20]

Jefferson wrote relatively little about the rationale behind his preference for a particular architectural style. But available evidence suggests that his affection for neoclassicism was based on three beliefs. First and foremost, he believed it was the paragon of architectural beauty. Second, because it had remained a bedrock of building design since its inception, Jefferson expected neoclassicism to remain attractive far into the future and thereby ensure that the appearance of the new capital would remain relevant. Third, Jefferson wrote that the best architecture should "improve the taste of my countrymen" and "increase their reputation" in the modern world, which neoclassicism presumably could accomplish.[21]

Finding Symbolism in Washington Architecture

Neither L'Enfant nor Jefferson suggested that symbolic language was a primary reason for embracing his preferred mode of architecture. However, this has not kept people from finding symbolic meanings in Washington's baroque map or in its neoclassical buildings, particularly meanings associated with the city's identity as the center of American government.[22] One common view is that they represent *republican government*. Looking at L'Enfant's map, for example, one finds parallels between the location of political institutions and their constitutional powers. The legislative and executive branches are separate yet connected; the Supreme Court (in what some believe was its intended location, between the White House and the Capitol) sits like a judge, balanced between the two other branches; and the higher altitude of Congress mirrors its political centrality.[23] Neoclassicism, meanwhile, is assumed to have been attractive to Jefferson and other Founding Fathers because it evoked the Roman Republic, which they saw as a model for the new government.[24]

A second widespread belief is that Washington architecture symbolizes *political power*. Baroque city planning, favored by many powerful European monarchs, can be ideal for "the staging of power"—witness the inaugural parade for new presidents along Pennsylvania Avenue. It also makes institutions of authority more visible: the Capitol dome can be seen along many avenues that depart from it "like gestures of command," as the architectural historian Spiro Kostof put it. The monumental and imposing nature of classical buildings also suggests power and strength, an association reinforced by their popularity with absolutist leaders like Napoleon and Mussolini.[25]

Finally, there are those who see not politics but *religious meaning* in the buildings and layout of Washington. One can find several neoclassical buildings on or near Capitol Hill that contain religious imagery. Many ancient classical structures, such as the Parthenon in Athens and the Pantheon in Rome, had spiritual purposes, leading some to describe Washington's neoclassical edifices as modern temples. President Herbert Hoover called the National Archives building a "temple of our history," for instance, while the central inscription of the Lincoln Memorial describes the building (fashioned after the Parthenon) as a temple.[26] The National Mall has been called "the great stage for American secular worship" and "a separate city of shrines," while Jeffrey Meyer writes that citizens make a "pilgrimage" to see Washington's "symmetries and axial boulevards, shrines, and monumental architecture whose underlying purpose is to give a transcendent meaning to the city."[27]

There is little evidence tying any of these meanings to L'Enfant or Jefferson, though subsequent city architects including Robert Mills did find architectural connections to broader values or symbolic ideals.[28] More importantly, these varying interpretations suggest the power of baroque city design and neoclassicism to convey symbolic meanings. As one scholar noted, "The baroque axis [of Washington] appears sufficiently flexible to represent any political system, from absolute monarchy to egalitarian democracy."[29] For instance, L'Enfant's layout could be seen as symbolizing the rule of the people because it features many views of *democratic* institutions. Another idea, reconciliation, is communicated by how the Arlington Memorial Bridge ties together memorials to Abraham Lincoln and Confederate General Robert E. Lee. How buildings are used can change their symbolic meaning too. For instance, the Lincoln Memorial has many details emblematic of post-war national unity (engraved names of the states, intertwined wreaths of Northern and Southern trees), yet it shifted from symbolizing national unification to representing individual freedom when it became a backdrop for major civil rights events.[30]

There is certainly nothing wrong with creative symbolic interpretations. After all, the Ancient Greeks and Romans preferred neoclassicism for reasons of their own: they believed that its symmetry and use of "ideal" numbers in measurements and ratios revealed eternal and universal truths and brought people closer to the natural order. Neoclassicism is no less powerful a style even though its original meaning has lost its significance.[31]

The ways that individuals interpret Washington's neoclassical buildings can have social or political consequences, however, especially if their interpretation conflicts with the buildings' surroundings. Before the Civil War, opposition to slavery in the city was exacerbated by descriptions of slave auctions right in sight of the Capitol building. The plight of impoverished black Washingtonians in the 1930s was made poignant by photos of impoverished alley dwellers living within blocks of neoclassical congressional office buildings.

Washington Architecture in the Nineteenth Century

Neoclassicism and baroque city planning may be powerful architectural styles, but whether the new capital would follow them over the long term was hardly guaranteed. Early Washington saw the construction of several major neoclassical edifices (or at least buildings with neoclassical elements), including the Capitol (started in 1793) and Washington City Hall (begun in 1820). However, as L'Enfant had acknowledged in 1789, "The means now within the power of the country are not such as to pursue the design to any great extant [sic]." And the city looked badly unfinished for many years.[32] Destruction by the British Army during the War of 1812 led Congress to consider abandoning the District altogether.[33] In 1842, the British novelist Charles Dickens famously called Washington a city of "magnificent intentions":

> Spacious avenues, that begin in nothing, and lead nowhere; streets, mile-long, that only want houses, roads, and inhabitants; public buildings that need but a public to be complete; and ornaments of great thoroughfares, which only lack great thoroughfares to ornament—are its leading features.[34]

Neoclassicism might have been abandoned entirely but for the Greek war for independence in the 1820s. Widespread compassion for Greece created a new wave of support for a more minimalist version of neoclassicism known as Greek Revival. In Washington this style was most famously employed by Robert Mills, designer of the Treasury Building (begun in 1836), the Patent

Office (also begun in 1836, together with another architect, William Elliot), and the General Post Office (started in 1839).[35]

By the 1840s and 1850s, the city was growing in size, parts of the city had begun to connect as L'Enfant had hoped, and the country's growing divisions over slavery had led leaders to encourage unity by improving the public areas of the capital. From the 1830s onward Washingtonians began calling their city a "metropolis," rather than merely the "seat of government."[36] Yet Washington still appeared incomplete. Emblematic was the Washington Monument, also designed by Mills. Its construction was halted in 1856 when its funding ran dry, and for decades the unfinished obelisk stood forlornly at the western edge of the Mall.

Furthermore, the development that did occur could depart rather starkly from L'Enfant's blueprint. Changes had appeared almost immediately after his map was made, when surveyor Andrew Ellicott modified his plan without L'Enfant's approval, straightening some streets, removing certain plazas, and shifting the location of the White House and Capitol buildings. Mills' Treasury Building blocked the reciprocal view of the White House and the Capitol; legend has it that President Jackson chose that site for the building because of his poor relations with Congress at the time.[37] But perhaps the most severe departure was along the city's main east-west axis. Intended as a grand avenue, the open area was slowly turning into a collection of individual parks, following to a limited extent a then-fashionable English-style garden design proposed by Andrew Jackson Downing in 1851. The red multi-turreted Smithsonian Castle (1855), which added to this more romantic image of the Mall, was built inside the area's intended border. The Washington Monument was also placed off-center from both main axes.[38]

In the early 1870s, post–Civil War economic expansion, the growth of a stronger national government, and leadership by the city's Board of Public Works—especially its vice president, Alexander "Boss" Shepherd, who later became territorial governor—brought civic improvements and new residents to the city. The Washington Monument was finally finished in 1884. The eastern side of the Potomac River was dredged, creating new land west and south of the Washington Monument. Memorials began to appear "as L'Enfant himself had imagined," and a seedy district known as Murder Bay southeast of the White House was in the process of being cleaned up.[39]

Nonetheless, while "the essential elements of the L'Enfant plan were retained," several new projects were undermining that plan. By 1899, a railroad station cut into the Mall, and the Capitol Building now had to share the skyline with the tall dome of a new Library of Congress building. Neoclassicism, furthermore, had fallen out of favor by the mid-1800s. Alternative styles, some quite dramatic, were being adopted for major city buildings. Besides the Romanesque Revival style of the Smithsonian Castle (see figure 1.2), they included Gothic Revival, French Second Empire (the ornate State, War,

Figure 1.2 The Smithsonian Castle
Photo by Matthew Green

and Navy Building, 1888), and Richardson Romanesque (the Post Office Building, 1899, which one senator called "a cross between a cathedral and a cotton mill"). And as the nineteenth century came to a close, new technology allowed the construction of even taller structures—starting with the twelve-story Cairo apartment building in 1894—that threatened to completely overwhelm Washington's existing skyline.[40]

Old Styles, New Symbolism: The McMillan Plan

The immediate danger of excessively tall buildings was avoided when the city's governing body, the Board of Commissioners, adopted a new limitation on building heights in 1894.[41] Later that decade, worried by Washington's physical condition, particularly on the Mall, local business interests and national architects suggested making enhancements in time for the 1900 centennial. They found a powerful (and, given Congress's authority over the city, important) ally in a U.S. senator, James McMillan of Michigan. Though too late for the centennial, McMillan persuaded Congress in 1901 to establish a commission that would recommend improvements to the city's parks and "incidentally suggest where the public buildings should be placed."[42]

The leader of the commission was the renowned architect Daniel Burnham. In 1893 Burnham had helped initiate a resurgence of the classical style in the United States as chief designer of the Chicago Fair, the "White City" of neoclassical structures that stunned and inspired visitors. Burnham, like L'Enfant, was a man of bold artistic perspectives, and his "powerful personality and expansive outlook dominated" the McMillan Commission, as the four-man group was commonly called.[43]

The commission's report, known as the McMillan Plan, declared neoclassicism to be the central feature of a revitalized Washington and "amounted to a firm endorsement and extension of L'Enfant's . . . baroque planning principles." It recommended revitalizing the core axes of the city, protecting long uninterrupted vistas in various locations, and ensuring that future structures would be neoclassical. The Plan sought to restore the Mall in particular, eliminating the train station, keeping all buildings equidistant from its center, and reimagining it not as a naturalistic and concrete "public grounds" but as a more abstract "public space" to be manipulated along rational lines.[44]

The commission's proposal proved wildly popular. This was in part because it was accessible to nonexperts, clearly written and dramatically illustrated. But it was also highly ambitious, proposing no less than to resculpt the core of a capital city with a boldness and grandeur that captured people's imaginations. And the plan's embrace of neoclassicism tapped into an emerging belief that the style symbolized civilized society and, importantly for a country slowly becoming a world power, national pride and strength. Its suggestion for a memorial to Lincoln was an appealing gesture of national unity and may have even ensured the plan's success—which was ironic, because the memorial's subject was not of great concern to the commissioners, who had added it primarily to complete the Mall's symmetry.[45] The McMillan Plan was so popular that it helped drive the so-called City Beautiful movement, a national push for civic improvements in America's growing, and often unsightly, cities—a movement which in turn created momentum to implement the plan in Washington.

Besides good prose, grandeur, and elements of implied symbolism, key to the plan's success was smart lobbying and politicking behind the scenes. Details of the report were given to the press in advance to encourage positive news coverage, and McMillan, his aide Charles Moore, and members of the commission carefully cultivated senators and cabinet officials to win their support. Once the plan's new vision for the city's center went public, its supporters, including Glenn Brown, the head of the American Institute of Architects (AIA), went on the lecture circuit and widely disseminated the plan.[46]

Implementing and Evaluating the McMillan Plan

The McMillan Plan inaugurated a period of intense construction in the city along neoclassical lines. New public buildings included the Central Library in Mt. Vernon Square (1903), Union Station (completed in 1908 and designed by Burnham himself), the city government building on Pennsylvania Avenue (1908), and the Lincoln Memorial (1922) (see figure 1.3). Edifices for private

Figure 1.3 The Lincoln Memorial
Photo by Matthew Green

organizations, such as the Daughters of the American Revolution (Memorial Continental Hall, 1911) and the Red Cross (1917), were also built in neoclassical form, as were several apartment buildings. These, along with the plan's inclusion of "national" monuments (such as the Lincoln and Grant Memorials), emphasized a view of the city as a symbol of national unity.[47]

Not everyone wanted to follow the McMillan Commission's directives, however. Just eighteen months after they were released, Burnham had to intervene to stop a new archives building from being placed near the White House, contrary to the McMillan Plan. When the architects of a new museum on the Mall proposed an ornamental design for its exterior, Burnham and fellow commissioner and architect Charles McKim forced a mid-construction change in its design to make the building more classically austere. Not long thereafter, an effort by the Department of Agriculture to impose its new headquarters into the Mall's open space was foiled only when AIA head Glenn Brown, along with members of the commission, organized a vigorous opposition campaign.[48]

This sort of ad hoc defense of the McMillan Plan was clearly not tenable over the long term. Accordingly, new agencies were given governing power over the city's development to ensure fealty to the McMillan and L'Enfant blueprints. The first was the Commission of Fine Arts (CFA), created in 1910. With Burnham and Moore among its first members (and Moore its chair from 1915 until the mid-1930s), it played a central role in dictating the style and location of public structures and won a critical early battle over whether the Lincoln Memorial would be a highway rather than a building. A second agency, the National Capital Park Commission, was established in 1924. Later reconstituted into a more powerful agency, the National Capital Park and Planning Commission, the commission published a series of plans for the city core in the ensuing decades intended to maintain a commitment to L'Enfant and the McMillan Commission. Both it and the CFA set rules for the size, appearance, and location of public buildings and forced architects to change their designs to fit the neoclassical model.[49]

Though the McMillan Plan rejuvenated L'Enfant's original map, not everything it envisioned was implemented.[50] Some have argued that it strayed from L'Enfant's schematic in troubling ways. River vistas and other views proposed by the French architect were blocked by the Lincoln Memorial and the Jefferson Memorial (1943), for instance, while the Supreme Court (1935) was placed not in Judiciary Square, as some believe he wanted, but behind the Capitol. Some of the fiercest condemnation of the plan has been over how it shaped the development of the Mall: not as a public promenade but what Lewis Mumford described as "a greenbelt, at best a fire barrier, which keeps segregated and apart areas that should in fact be more closely joined."

Based on these and other departures from L'Enfant's plan, one influential critic concluded: "I think L'Enfant would say that *his* city of Washington was never built."[51]

On the other hand, the McMillan Plan may well have prevented more drastic deviations from the overall baroque and neoclassical styles, particularly with the rise of austere modernist architecture and greater pressure to build freeways through the city in the mid-twentieth century. The commission replaced the haphazard method of locating buildings with a single vision of how each structure fit into the city as a whole. Some even applaud its reimagining of L'Enfant's city, including the Mall. The latter has become a place of protest, a "symbolic confrontation between the people and their representatives," wrote one observer, and its museums and monuments have established "the Mall as historical symbol" and "the physical representative of American history as an ideal." Though perhaps less oriented to daily city life than L'Enfant wanted, the Mall's design means that, in the words of architectural historian Witold Rybczynski, it "belongs to the entire continent."[52]

Conclusion

"In cities," wrote Spiro Kostof, "only change endures." Washington is no exception. Shifting needs and architectural styles have introduced considerable architectural variety in Washington. As the city grew beyond its core, streets emerged that curved, bent, or twisted away from the geometric layout developed by L'Enfant. Major nonclassical buildings were added in the latter half of the nineteenth century, and again after World War II. Modern structures like I. M. Pei's triangular East Gallery and the curving Museum of the American Indian, both on the Mall, challenge the neoclassical uniformity desired by the McMillan Commission and its like-minded successors. Other famous modernist architects are represented in the city by such buildings as the Kreeger Museum (by Philip Johnson) and the Martin Luther King Jr. Memorial Library (by Mies van der Rohe). Even Daniel Burnham, who possessed the "bulldozing habit of mind" common to baroque city planners, could not get "nonconforming" buildings like the Library of Congress Building, the Arts and Industries Building, or the Smithsonian Castle moved or demolished.[53]

Still, amid this architectural diversity, L'Enfant's map remains a sacrosanct blueprint for Washington, and neoclassicism reigns as the principal style of the capital's public architecture—thanks to both its endurance and the periodic support of Congress and influential architects. In 1888, Congress mandated that all future suburban development within the district be laid out in conformity with L'Enfant's street grid to the fullest extent possible. In the 1960s, when neoclassicism was shunned by most conventional architects, the subway stations for the future Washington Metro were designed to echo classical archways and ceilings.[54] And neoclassical elements continue to be seen in such recent edifices as the Ronald Reagan Building (1998) and the World War II Memorial (2004).

Those visiting Washington today are offered striking views of remarkable monuments and buildings. The size and limited ornamentation of the city's neoclassical edifices convey majesty and authority, and the visual similarity of its structures with those from the ancient world creates a sense of timelessness. As a result, even if "we can no longer read the language of classicism," as Nathan Glazer argues, Washington's great buildings, and the baroque arrangement of its streets and public spaces, continue to evoke awe and appreciation among those who live, work, and visit the nation's capital. And if "the architecture and design of a capital city has as its principal function to *create citizens*," in the words of political scientist Margaret Farrar, the boldness of Washington's architecture would certainly seem to fulfill that function.[55]

CHAPTER 2

Memorialization, the Mall, and the National Imagination

As Pierre L'Enfant surveyed the territory of the newly minted capital in 1791, he imagined a city filled with wide boulevards and monuments befitting what would become a great empire. While the city would fall short of such expectations for its first several decades, it eventually acquired the status its founders had envisioned for it, though in a modernist mold. Today Washington contains hundreds of monuments and memorials, serving as coveted territory for those seeking to commemorate a person, event, or cause on a national scale. As a centrally planned, symbolically important, and highly contested ground for commemoration, the Mall and its structures comprise a nationally significant locus of city culture.

A two-mile long tract located in the western part of Washington, D.C., the National Mall is home to the "monumental core" of the city, forming the central axis upon which many of the most recognizable monuments and memorials are constructed. Considered the most treasured block of urban public land in the United States, the Mall has been called "America's backyard" and is widely regarded as a sacred national space. Most of the millions of tourists to the city consider a visit to the Mall's monuments and memorials the highlight of any Washington, D.C. pilgrimage.[1]

Washington, D.C.'s central monuments and memorials are, at their most basic, commemorative structures that function within the context of the unique national space of the Mall. A shift in the way the Mall's open space was perceived in the early 1900s led to a greater emphasis on large-scale planning and architectural relationships between commemorative structures on the Mall. Moreover, changes in commemorative culture, design, and practice have shaped our understandings of the meanings of the memorials and the monuments over time. And the Mall's commemorative structures and the events that take place around them illuminate public attitudes toward the nation.[2]

Memorial vs. Monument

Washington is full of memorials and monuments, but what is the difference, if any, between the two? A *memorial* is defined as "something, such as a monument or holiday, intended to celebrate or honor the memory of a person or event"; a *monument*, as a "structure, such as a building or sculpture, erected as a memorial." The standard definitions of the terms overlap: both are intended to commemorate and/or remember one or more persons or events. However, while a monument can be celebratory and tends toward veneration, memorials are more likely to evoke somber remembrance. Perhaps because of the association of "memorial" with solemnity, new commemorative structures are more often called "memorials" than "monuments," though, as we will see, several recently built structures designated as memorials have the celebratory elements of monuments.[3]

Two commemorative structures on the Washington Mall illustrate the difference. The Vietnam Veterans Memorial Wall, or the "Vietnam Wall" as it is popularly known, has sunken granite walls engraved with the names of the more than 58,000 servicemen who died or have been designated missing in action in the conflict (see figure 2.1). It is a deliberate rejection of the traditional monument: its reflective surface intentionally punts interpretation of that war back to the viewer and eschews celebration. The architect, Maya Lin, even referred to her design as "antimonument." By contrast, the far more assertive Washington Monument, a 555-foot obelisk prominently located near the center of the Mall and the tallest structure in the city, is intended to evoke power. With its Egyptian design, soaring pyramidion, and prominent position, the monument symbolically links the nation's first president to a powerful ancient civilization and is clearly celebratory. Upon its dedication in 1885, *The New York Times* reported that the Washington Monument had become a symbol for "a new era of hope"—whereas one observer complained that the Vietnam Memorial's design was not "inspirational like the Washington Monument."[4]

Memorials and monuments can differ dramatically in appearance as well. The traditional monument is typically a unitary physical structure, often set on a pedestal, that mediates its site as well as the symbolic power it possesses. But a memorial can be a park, museum, nature preserve, library, or any variety of temporary, spontaneous commemoration, and the term *memorial* incorporates the wider range of viewpoints on the meanings that memorialization holds for Americans today.[5]

Figure 2.1 The Vietnam Veterans Memorial Wall
Photo by Matthew Green

Some argue that memorials are preferred over monuments and are increasingly popular in the United States, including Washington, D.C. In this view, memorials are reflective of the multiple beliefs, perspectives, and cultural values of the times in which they are created. The antiwar movement of the 1960s, for example, informed the design of the Vietnam

Veterans Memorial Wall. The revelations of how the Vietnam War was waged in the 1970s, the failure to achieve the goals set out for that war, and the ambivalences of the veterans' experiences during and after the conflict informed Maya Lin's designation of her memorial design as an antimonument.[6]

Memorials, Monuments, and the Power of Commemorative Experience

The feeling and sense of meaning a memorial or monument evokes is, in any case, tied to each individual's experience of the structure. While the range of such experiences is wide, historian John Bodnar offers two useful categories for their analysis and how they contribute to public memory: official expression and vernacular expression. Cultural authorities, such as political leaders, are more likely to use official expression in commemorations, which stress social unity and loyalty to the status quo and existing institutions. Vernacular expression, however, is more reflective of the multiple perspectives and interests within broader society. Such expressions tend to be more diverse, can clash with each other, and can be formed and reformed over time. Vernacular expression, moreover, often threatens the official versions of memorialization that serve existing power structures.[7] Official literature and ceremonies held at Washington, D.C.'s memorials and monuments (such as dedication ceremonies and other state-sanctioned events) usually employ official expression, conveying a very different meaning than do demonstrations and rallies that are not sponsored by the state. Such ceremonies can be organized and attended by different groups of people, hold different purposes, and generate contrasting views of the nation.

Aside from the language used to describe them, memorials, like any architectural work, can mobilize audiences to great emotional depths. Imagery depicted in memorials can achieve this effect, particularly if those images come to express strongly held views about the nation. A memorial image can be a concrete representation of a particular person or event: the nineteen-foot statue of Abraham Lincoln in the Lincoln Memorial, for example, or the depiction of marching Union soldiers on the Ulysses Grant Memorial located at the base of the U.S. Capitol Building. Symbolic images are also effective conveyors of meaning. One of the most recognizable is the bald eagle, which represents the nation and adorns countless memorials. And some memorials become so iconic—such as the Lincoln Memorial—that they become symbols in themselves, as noted in the previous chapter.

Memorials, Monuments, and Changing Views of Space

Most people assume that the National Mall is the natural site for the nation's memorials and monuments. However, the history of its development and planning reveals how changing conceptions of space shaped the ways individuals saw commemorative structures as fitting—or not fitting—within it.

Pierre L'Enfant anticipated the construction of monuments and memorials in Washington, proposing many squares and plazas that might house them. The Mall itself was planned as a grand walkway extending about a mile from Capitol Hill, bordered by museums and theaters. Like the capital itself, however, the Mall area had been a victim of unmet expectations throughout its earliest decades. In 1804, visiting Irish poet Thomas Moore penned his thoughts on the "embryo capital" in verse:

> Come let me lead thee o'er this second Rome . . .
> This embryo capital, where Fancy sees
> Squares in morasses, obelisks in trees'
> Which second-sighted seers, ev'n now adorn,
> With shrines unbuilt and heroes yet unborn . . .[8]

James Sterling Young, historian of early Washington, noted that "where monuments had been planned, brush piles moldered and rubbish heaps accumulated."[9]

Though development did eventually come to Washington, post–Civil War design trends pushed the memorial landscape further from L'Enfant's original plan. As historian Jon Peterson notes, "The neoclassical design principles of harmony and regularity that had guided [L'Enfant] had given way to ones predicated on irregularity and intricacy derived from medieval art." New architectural tastes of the Victorian era were responsible, but so too was the absence of a political and economic commitment to an orchestrated design for the Mall. Instead of fulfilling L'Enfant's classical vision for a unified monumental core, the space became a disconnected collection of parks and buildings. The Tiber Canal, which Andrew Jackson called a "dirty, stinking, filthy ditch" because it served as both an open sewer and a storm drain, lent itself to a general air of unsanitary neglect. Architectural historian Richard Guy Wilson calls the nineteenth-century Mall "an unkempt gardenesque park, with no particular symbolic value."[10]

Indeed, before the city's monumental core was developed, a kind of "statue mania" would take hold in the city, though not on the National Mall. The equestrian statue, a heroic depiction of a prominent individual mounted on a horse, was intended to offer citizens an image of a hero whose patriotism could be emulated. This type of "heroic statue" originated in the ancient world and was an extremely popular commemorative form in the nineteenth-century United States. Indeed, L'Enfant had originally planned that an equestrian statue of George Washington be placed in the location now occupied by the Washington Monument.[11]

Clark Mills' 1853 equestrian statue of Andrew Jackson in the city's Lafayette Square was the first of such structures in the nation. Its construction helped trigger an equestrian statue craze, and these free-standing sculptures, primarily of soldiers, were built all over Washington. They included likenesses of President Ulysses Grant and Generals Nathaniel Greene, George McClellan, James McPherson, Winfield Scott, William Tecumseh Sherman, and George H. Thomas. The equestrian statue of George Washington envisioned by L'Enfant was eventually built in 1860, not on the Mall, but in Washington Circle, by Mills. Washington, D.C. was one of the largest repositories of these sculptures in the world by the early twentieth century.[12]

As the statue craze grew with the demand for civic art in the nineteenth century, however, a backlash followed. Such statues came to be viewed as dull, stiff, and unimaginative and were criticized for their placement in the city's public squares and circles without attention to broader landscape and design considerations. As one critic put it, the statues were "inserted as independent objects, valued for their historical or memorial qualities or sometimes for their individual beauty, regardless of their effect on their surroundings." A newspaper columnist opined in 1901 that the city's statue monuments were "perhaps the most hideous in the world."[13]

In the meantime, the Mall saw haphazard development. Architect James Renwick was asked to design the Smithsonian Institution's new quarters in the 1840s. The Norman-style castle, regarded as one of the great works of nineteenth-century U.S. architecture, was the result, though the structure is a departure from the neoclassical style of many of the Mall's other edifices (see figure 1.2). With the construction of "the Castle," Andrew Jackson Downing, the most well-known landscape planner of the mid-nineteenth century, was asked to design a plan for the Mall grounds surrounding the new building. He proposed a collection of parks for the area in 1851, including a "monument park" with trees and vistas intended to instill civic virtue. Downing's plan broke with L'Enfant's and set the stage for the more leafy, if piecemeal, development of the Mall discussed above.[14]

The Washington Monument was also constructed in the nineteenth century, its elegant simplicity belying its complex and tangled history. Indeed, architect Robert Mills' original design for the structure featured a colonnade building 250 feet in diameter with 30 columns, each 45 feet high; every state's coat of arms was to be inscribed above the columns, and a pantheon of Revolutionary War heroes would be featured inside the colonnade building.

From this ornate structure would arise an obelisk, the only feature of the original design in evidence today. The cornerstone for the monument was laid in 1848, before a crowd of 20,000. In 1852, however, the project was suspended due to lack of funds, leaving an unfinished shaft 170 feet high.[15] An 1877 letter to the *Washington Post* complained:

> Near the filthy marsh in the rear of the White House . . . stands an unsightly stone structure called by jobbers a "Monument to the Immortal Washington." For thirty years it has stood there, a disgrace to the name of the father of his country.[16]

The monument was finally completed in 1884.[17]

By the late nineteenth century, an emphasis on city improvement led to efforts to create a more attractive city, particularly the Mall, culminating in the McMillan Plan's highly organized monumental core. But the new plan was more than just a vision for where buildings should go. It also reflected, according to Kirk Savage, a move from "the nineteenth-century concept of *public grounds* to the twentieth-century concept of *public space*." The public ground, he argues, is a place perceived as "concrete, tangible, messy, [and] diverse," where buildings or memorials may be placed as "independent objects." This was L'Enfant's view of the Mall, and it was a common view of public areas by planners and landscape architects of the 1800s. The McMillan Commission, by contrast, saw the Mall as public space where there "is no longer mere emptiness or the enchanted realm of God but a medium that human beings now claim the power to control and manipulate. With the aid of human ingenuity, space can now flow, envelop, expand, or contract." When the older system of design was abandoned, the ground became a platform for implementation of spatial effects and could be subjected to modern ideas of control and design. In short, the new planners saw the redesigned Mall's memorial and monument structures more in terms of "spatial ensembles" than "independent objects."[18]

Though the McMillan Commission and its leader, Daniel Burnham, were willing to reshape the Mall area to comply with their new vision, they also sought to make many existing monuments and edifices fit within the revitalized Mall, reconfiguring them as parts of the Mall as a whole. The Washington Monument, dedicated in 1885, figured hugely in the new plans. Viewed as an "eccentricity" by landscape architects at first, ideas about the monument shifted by 1901, when those tasked with the Mall's redesign began to see it as the centerpiece of an expanded space. As Savage notes, what had been ridiculed as a "chimney" and a "mechanical monster" was now seen as an expression of the power of nature and the ingenuity of man. The space around the enormous obelisk was regarded as integral to the enhancement of the entire Mall. The monument, writes Savage, "was now the fixed, unchanging point around which [the Commission members] wove their dreams of a new monumental core." In the new plan, a Union Square was placed at one end, and the other side featured an extension of the Mall, a memorial to Lincoln, with the obelisk framed in the center. "The plan," as Savage notes, "enshrined a national narrative of reunification" that the commission members "believed would be universal and timeless." The monumental core could then "provide a transcendental experience of the federal nation-state, a way to understand its spread and its authority not as brute physical power but as a work of an irresistible and beautiful system."[19]

This model, which demanded tremendous resources and discipline to implement, is not without its critics. Michael Lewis points out that "unlike the permissive Mall of the nineteenth century, which could tolerate a considerable amount of experimentation and change, the McMillan Plan brought with it the frosty and unforgiving nature of a complete work of art."[20] Nonetheless, it set the stage for the next century of development.

The redesigned Mall, like the idea of America itself, was neither universal nor timeless, whether the McMillan commissioners believed so or not. Shifting ideas about the nation, however, worked in conjunction with shifting concepts of space and commemoration at the turn of the century to generate a transformation of the National Mall and its meaning to Americans.

Changing Styles and Meanings: Three Examples

Several important themes emerge from a careful examination of Washington monuments and memorials. To illustrate those themes are three important examples from very different eras of monument construction: the 1860s *Freedom* statue atop the Capitol dome, the Lincoln Memorial of the 1920s, and the Vietnam Veterans Memorial of the 1980s.

Freedom

The founders of the nation and of the city itself held lofty ideals as to what the nation meant and what the U.S. capital and its monumental structures should look like. L'Enfant biographer Scott Berg notes that "the entire city was built around the idea that every citizen was equally important," adding that the "Mall was designed as open to all comers, which would have been unheard of in France. It's a very sort of egalitarian idea."[21]

That "every citizen was equally important" was, of course, a fantasy. The American ideal of egalitarianism was far from a reality in L'Enfant's time. The inequities as well as the ideals were evident in the nineteenth-century Mall's commemorative structures. As Savage notes, "From the start, the city's claim to represent an 'empire of liberty' clashed with its location in the very cradle of American slavery."[22] The land itself had been seized from its original native inhabitants, who were duly excluded from the benefits of U.S. citizenship. This did not prevent the earliest political representatives and their statuary designers from embedding such ideals in their work, though the contradictions embedded in this idealized imagery came with the territory.

One of the Mall's earliest statues, *Freedom* offers an example of how ideals and their contradictory elements can be ensconced in the images, symbols, and reception of the Mall's memorials and monuments. Set atop the U.S. Capitol Building, *Freedom* is a nineteen-foot bronze figure of a woman holding a sheathed sword in her right hand, a shield and wreath in her left. The statue's headdress, often mistaken for a Native American style, is a helmet ringed by stars bearing the head of an eagle.

Despite the theme of the statue, the design and construction of the monument underscored the hypocritical presence of slavery in the American republic, perhaps more so than any other monument in the city. Originally fashioned wearing a liberty cap, similar to those worn by freed Roman slaves, sculptor Thomas Crawford was forced to redesign the statue's headwear when Secretary of War Jefferson Davis objected that it smacked of abolitionist propaganda. When *Freedom* found her way to the Capitol's tip in 1863, Davis was busy leading the charge against Union forces as president of the Confederacy.[23]

The Lincoln Memorial

At the other end of the Mall from the Capitol sits the Lincoln Memorial (figure 1.3). Dedicated in 1922, the structure features elements that point to Lincoln's role in ending slavery, such as Jules Guerin's mural of an angel liberating a slave.[24] The memorial is among the most famous in Washington. It is also the quintessential example of how the meaning of memorials can change with the shifting values and beliefs of the nation as a whole.

Originally intended to sit at the current location of the Jefferson Memorial, at the south end of the White House axis, the Lincoln Memorial site was moved to the opposite end of the Capitol amid design disagreements.[25] The Commission of Fine Arts, the group charged with the implementation of the McMillan Plan, set several criteria for the design and construction of the memorial, illustrative of the emerging concepts of space and planning outlined above. These included the following:

- The structure had to complement the existing structures on the Mall, specifically the U.S. Capitol and the Washington Monument;
- The structure had to depict Lincoln as a savior of the nation who was nonetheless, as Christopher Thomas notes, "a modest man of no pretension";[26]
- The memorial would serve as the primary memorial to the Civil War in the city; and
- The structure had to set, as Thomas puts it, a "reflective, elegiac, and compassionate tone, like the remembered Lincoln himself."[27]

The criteria marked a departure from the Victorian nineteenth-century Mall development from the grounds-centered concept to the Mall as a more sweeping and unified public space. It also led to a design duel between two architects, John Russell Pope and Henry Bacon, with Bacon receiving the commission to build the current structure.[28]

The result was as the commission wished: a neoclassical structure that harmonizes with the other two main structures on the central Mall axis in a modern sense of space and landscape planning. Modeled after the Parthenon in Athens, Greece, its flat top complements the domed Capitol building and brooks no competition with the Washington Monument. Dedicated on Memorial Day in 1922, the memorial was instantly popular, seeing a third to half a million visitors in the year after its construction.[29]

In the case of the Lincoln Memorial, the site became a place for large public events—rituals that celebrated "collective, usually national values, said to be shared by all who make up the collectivity."[30] The first such event, the memorial's dedication, conveyed the official version of the memorial's intent: to represent national unity after the divisiveness of the Civil War. Authorities could easily point to features of the memorial's design to support this interpretation, such as the wreaths made of enjoined northern laurel and southern pine and inscriptions emphasizing union. The planners also believed that the structure would be used mostly for private visiting and contemplation, not public rallies.[31]

Vernacular cultural expressions would hold a different, if overlapping, set of meanings for the memorial than official expression made by the dedication ceremony. By the late 1920s, with the permanent lighting of the memorial for evening visitation, the growth of tourism, and the increasing use of cars, the memorial became a site for large numbers of visitors. The neoclassical memorial also became a locus of modern ritual. The year 1940 saw 1.5 million visitors to the memorial; by 1950, 2.5 million visitors were making pilgrimages to the memorial; the 1970s and 1980s saw between 2.5 and 4.5 million per year, and the annual number hovers around 3.5 million today.[32]

A more public use for the memorial was, in fact, considered as early as its dedication ceremony. A *Washington Post* reporter claimed at the time that the setting had a potential "for national fetes and spectacles," with the "noble colonnade as its background."[33] Scores of groups would stage demonstrations, rallies, and commemorative events at the memorial, but it was the African American civil rights movement that made extremely effective use of the structure and its grounds to illuminate their status as second-class citizens. It began in 1939 when Marian Anderson, a world-renowned black opera singer, gave an outdoor concert at the memorial after being denied the right to perform at the Daughters of the American Revolution's (DAR) Constitution Hall.[34]

Historian Scott Sandage has shown how African American activists used the space to make racial inequality a national issue between that Easter Sunday concert and the 1963 March on Washington. "A standardized civil rights protest ritual evolved," he writes, that used "mass rallies instead of pickets, performing patriotic and spiritual music, choosing a religious format, inviting prominent platform guests, self-policing the crowds to project an orderly image, alluding to Lincoln in publicity and oratory, and insisting on using the memorial rather than another site." In 1947, when Harry Truman became the first president in history to give a speech at the annual meeting of the National Association for the Advancement of Colored People (NAACP), he did so at the Lincoln Memorial. Twenty-six years later, during the March on Washington for Jobs and Freedom, the connection between the memorial and civil rights was made concrete to millions of Americans as Martin Luther King, Jr. was televised delivering his famous "I Have a Dream" speech with the memorial in the background.[35]

Memorials, then, are important sites of memory where ritual can be used to transform the meanings of national symbols. Particularly in times of great national transformation, protesters can mobilize mainstream symbols such as the Lincoln Memorial "to further alternative ends, to constitute (not just reflect) shared beliefs, and to open spaces for social change." As Savage notes, the Anderson concert of 1939 established the Mall as a place of "moral principle" defined by "the citizens who occupied it . . . a so-called minority group had shown Americans how to fulfill their democratic potential in the nation's public space."[36]

The Vietnam Veterans Memorial Wall

The Vietnam Veterans Memorial presents another example of the interactions between official and vernacular expressions and how they are negotiated in public space. Unlike the Washington Monument and the Lincoln Memorial, the Vietnam Veterans Memorial commemorates those who were killed, missing, or served in a war, and it was designed and constructed while many of the war's veterans, and members of the families of those killed, were still alive. Proposed by Vietnam War veterans Jan Scruggs, Tom Carhart, and other veterans in the Washington, D.C. area in 1979, the memorial also became the focus of intense conflict over the meaning of memorial commemoration.

The context of the Vietnam Wall's construction was as different from memorials that came before as that of the Lincoln Memorial's to the Washington Monument's. By the time the Vietnam Veterans Wall was proposed, the McMillan Plan's implementation was complete. The Lincoln Memorial, the Washington Monument, and the U.S. Capitol formed the anchors that defined the main axis' parameters, and the Jefferson Memorial, dedicated in 1943, completed a north-south axis from the White House to the Tidal Basin of the Potomac. Museums now lined the Mall and offered world-class cultural institutions to complement the monumental core (see chapter 3). And the United States, although humbled by the social upheaval of the sixties and the losses of the Vietnam War, was nonetheless still recognized as a global superpower.

The conflict over the Vietnam Veterans Memorial was in part a product of ambivalence over the meaning of the war itself. Scruggs, for example, stressed that the new memorial should promote healing and reconciliation. Government officials involved in the planning, however, emphasized national sacrifice and loyalty. The location near the Lincoln Memorial was selected because that memorial exemplified the unity of the nation, and the hope was that the new structure would contribute to that unity. Maya Lin's final design, however, reflected grief and loss more than national celebration and glory. Its two long, black granite walls bearing the names of each of the more than 58,000 individuals who died in Southeast Asia between 1957 and 1975 hardly symbolized national strength. The design was characterized as depicting "a mass grave," and Carhart, initially a project promoter, called it a "black gash of shame," arguing instead for "something that will make us part of America." In the wake of such protests and in spite of Lin's objections, a decision was made to add an American flag and a "heroic statue" featuring three soldiers from the conflict and to inscribe "God Bless America" on the statue itself.[37] The original design would stand, however.

The Vietnam Veterans Memorial changed the nature of national commemoration starting with its official dedication in November 1982. Rather than solely expressing patriotic sentiment celebratory of the nation, the dedication ceremony included expressions of appreciation for ordinary soldiers and grief over the loss of loved ones and fallen comrades. Marching in the ceremonial parade were soldiers in and out of uniform, and placards critical of the engagement were as much in evidence as those commemorating it—a far cry from the traditional celebratory and orderly memorial dedication parade, and certainly very different from the dedication of the Lincoln Memorial decades earlier. Those who visited the wall were moved to tears, and expressions of anguish occurred from the very beginning. The practice of leaving objects at the memorial, now routine, began when the parents of one dead soldier left a pair of cowboy boots there just after the dedication. By 1985, thousands of objects left at the wall were being collected by the National Park Service.[38]

This memorial, more than any in the city, became a site of personal commemoration and expression of loss, where viewers read themselves into the structure and could complete its meaning by leaving personal objects. It also reinvigorated the American tradition of memorial building. The wall served as the first true memorial to victims that existed, as Savage notes, "not to glorify the nation but to help its suffering soldiers heal—Maya Lin's model bequeathed to us a therapeutic model of commemoration" that is now the core model of memorialization in the United States.[39] It also raised questions as to whether other veterans were being properly memorialized in the United States and on the Mall. The result was the building of more war memorials in Washington: the Vietnam Women's Memorial in 1993, the Korean War Veterans Memorial in 1995, and the National World War II Memorial in 2004.

The Vietnam Veterans Memorial's minimalist style, and its emphasis on individuals and groups over national themes, also became the standard for future memorials in Washington and elsewhere. One of many examples is the African American Civil War Museum and Memorial, dedicated in 1998, which follows the pattern of the Vietnam Veterans Memorial Wall, though it is set in the historically African American U Street neighborhood. The memorial features an eleven-foot statue of six soldiers by the sculptor Ed Hamilton and is surrounded by granite walls featuring the names of over 200,000 black soldiers who fought for the Union during the Civil War. Hence this off-Mall national memorial features some traditional monument features, yet the names inscribed on black granite echo Lin's antimonument aspect as well.[40]

Not all memorials have followed Maya Lin's lead. The largest recent example of a contrary style is the National World War II Memorial, set in a prime location between the Lincoln Memorial and Washington Monument. At the time of the memorial's construction, veterans of WWII had been commemorated abundantly, on a local and national level. War materiel was salvaged and used in tourist attractions, weaponry was displayed in parks and public buildings, and many living memorials—parks and museums—were created in the 1940s, 1950s, and 1960s.[41]

Nonetheless, many claimed that the memorial was needed because veterans of the Second World War had never been properly memorialized. Historian Erika Doss asserts that building the World War II Memorial had less to do with the absence of memorials than with the creation of social consensus and political obligations related to a current culture of gratitude. Such expressions often follow periods of social conflict and, in this case, "reinforce assumptions of America's historic and habitual militarism."[42]

At 7.4 acres and bearing a $175 million price tag, this was indeed a stupendous project. Composed of "white granite instead of black, plaza instead of park, loud instead of hushed, overflowing with words instead of stripped down and minimalist," as Savage puts it, the World War II Memorial looks more like a rejoinder to Lin's structure than a descendant.[43] Perhaps the fact that the World War II structure has the personality of a monument rather the more somber air of the Lin-style memorial is shaped in large part by the fact that the personality of the nation shifted into a defensive, militarist mode after the September 11, 2001, terrorist attacks, though plans for the structure were well under way by then.

Conclusion

The National Mall and its commemorative structures were initiated haltingly, in fits and starts, starting with L'Enfant's plans for a set of cultural institutions dotted with monuments to national heroes in the eighteenth century. The slow growth of the capital prevented the implementation of his plans in the nineteenth century, with the prevalence of Victorian ideals and design models governing Mall plans until the McMillan Commission resuscitated many of L'Enfant's neoclassical ideals in the context of new ideas about space and nation in the twentieth century. From an emphasis on traditional commemorative structures bearing celebratory features, best exhibited in the soaring Washington Monument, commemoration took a more somber tone, exemplified by the Vietnam Veterans Memorial Wall. The need to emphasize the more celebratory aspects of the nation never disappeared, however, as suggested by the placement of the soldiers' statue near the Vietnam Wall and the design of the World War II Memorial, which appears to move memorial design back into a more celebratory, commemorative mode.

But design and meaning will continue to change. Indeed, the latest addition to the Mall's monumental structures presents a milestone in commemorative structures. Dedicated in October 2011, the Martin Luther King, Jr. Memorial, set on the edge of the Tidal Basin, is a memorial honoring an African American pacifist who was not a military figure, nor a prominent civic official, nor an official holder of public office. The King Memorial features a

Figure 2.2 The Martin Luther King, Jr. Memorial
Photo by Matthew Green

thirty-foot tall likeness of the reverend emerging from a mountain of granite. The design is taken from a quote from King's "I Have a Dream" Speech, delivered on the Mall in 1963: "Out of a mountain of despair, a stone of hope" (see figure 2.2). As critic Edward Rothstein notes, while the image of King is taken from a photograph, the stone image is less a thoughtful than a determined King, and that it is "monumental, not human."[44] This may be the case. But a memorial in the monumental mold to a figure such as King, one who presents the inversion of the traditional Mall hero, is a novelty for "America's backyard." And however it may be interpreted today, it is likely to carry different—and possibly unintended—meanings in the future.

CHAPTER 3

A City of Magnificent Museums

Washington, D.C., is a city known for its museums. Drawing visitors from around the nation and the globe as well as local residents, the capital's museums are among the most visited in the world.[1] But in addition to its most popular sites—the Air and Space Museum, the Museum of Natural History, the National Gallery, and others located on the National Mall—there are many, many others. Seventeen Smithsonian Institution museums and art galleries, over forty historic houses, and dozens of other public and private museums on topics as diverse as espionage, medical research, and postage stamps are located in the Washington, D.C., metropolitan area.[2]

The average gallery visitor probably gives little thought to what she sees or why it is there. But all museums confront complex challenges as they strive to collect, exhibit, and interpret information amid changing technology and visitors' expectations and needs.

In addition, museums in the District of Columbia face unique challenges. The city's public facilities face the practical problem of operating with federal funds in a city governed by Congress, and they must carefully negotiate their relationship with the federal government—especially if they choose to display material that is politically charged. Those seeking to create a new museum with public money must be especially adept in such relations, particularly since questions about if and where museums can be built and how they are funded can be thorny. In addition, because they are located in the national capital, museums in Washington are reflections of national identity, and in many ways they are expected to work together to tell a national story. Telling that story involves difficult tasks of defining what it means to be an American and constructing public memory.

The Purpose and Role of a Museum

In the nineteenth century, museums were understood to be simply repositories of artistic, scientific, or historical artifacts, and their main purpose was to store and exhibit these artifacts. Though this is still central to a museum's mission, that mission has also grown more complex, especially for preeminent museums. Conservation techniques have become increasingly sophisticated. Scholars from around the world visit museums and their collections to conduct research. Above all, because they seek to educate as well as entertain and because they face competition with many forms of entertainment, museums cannot simply put things on display thoughtlessly. Museum staffs recognize that visitors must somehow be engaged during their visit in order to connect to both the physical space and to the content included in the exhibits. If this occurs, they are more likely to encourage others to visit the museum and to come back themselves.[3]

This is especially true in Washington, D.C., home not only to some of the preeminent museums of the country but also to many tourist attractions that all compete with each other for a limited audience. Here one can find, for instance, examples of the most radical version of visitor engagement: including visitors in the process of exhibit creation.[4] The American Art Museum has a program called "Fill the Gap" in which visitors are asked to select the best artwork to display (see textbox 1). Other museums in Washington seek to emulate the model of the amusement park. Visitors to the Spy Museum, for example, are assigned a secret identity to adopt and memorize as if they are on a covert mission. A more grim example of such visitor engagement can be found at the U.S. Holocaust Memorial Museum, where people are given identification cards that tell the story of a survivor or victim of the Holocaust, which they read while being transported in an ominous steel elevator car to the museum's main exhibition.

TEXTBOX 1: The Visitor Experience at the Luce Foundation Center for American Art

The Luce Center, housed in the Smithsonian American Art Museum, is an example of how D.C. museums invite visitors to experience art in a variety of interesting and innovative ways. First opened in 2006, the Luce Center's primary purpose is to store art from the museum's collections that is not currently on display. Instead of hiding all of these pieces, the Luce Center is equipped with hundreds of glass cases and storage drawers to make the art accessible to the public and allow individuals the opportunity to experience and respond to the art in a personal way.

The fact that the site is not a traditional gallery creates an opportunity for experimentation in programming and innovation in outreach and exhibitions. The collections are displayed in visible cases with very little physical interpretation. Art is clustered together in four broad categories: paintings, sculptures, folk arts, and crafts. Visitors are invited to write down accession numbers and look up more information about pieces that interest them on a website or at information kiosks located throughout the space. Visitors can also listen to information available at "audio stops" throughout the space.[5]

The Luce Center has also worked to expand its visitor base through innovative programming and to allow both physical and virtual visitors to contribute to the museum in a variety of ways. For instance, the center provides opportunities for visitors to play an active part in content decisions for the museum. Its Fill the Gap program allows visitors to choose artwork to be displayed in empty spaces in the display cases, and the museum utilized input and feedback from potential visitors to design the Art of the Videogame exhibit.[6]

There are more subtle ways that museums can enhance visitor experience and further their educational mission, of course. But nearly all of them make some use of technology. Sound and lighting can be employed to create a sensory experience. Videos can be incorporated into exhibits. Computers can be utilized in a variety of ways within the physical space. The Internet is also increasingly being used to complement physical exhibitions. As the public has grown accustomed to personalized, customized, and on-demand experiences that are easy to access and simple to share and build upon, museums have to consider how they incorporate personal levels of interactivity to tailor their programs and information to individual needs.[7] Again, examples of this can be found in many Washington museums. For instance, at the Lincoln Cottage museum in northern Washington, one exhibit includes computer displays allowing visitors to simulate being a cabinet member in President Lincoln's White House, while another lets visitors decorate their own video image with the iconic clothing of the former president (see textbox 2).

TEXTBOX 2: Technology at Lincoln's Cottage at Soldiers' Home

Technology plays a significant role in the exhibit space at Lincoln's Cottage at Soldiers' Home. This site, which is maintained by the National Trust for Historic Preservation, opened in 2008. The exhibits in the education center provide valuable background information about Lincoln, his presidency, and his connection to the cottage at Soldiers' Home. Two of the permanent exhibit rooms have large central screens playing videos on continuous loops. In the Lincoln's Toughest Decisions gallery, visitors become part of Lincoln's Cabinet and are asked to advise him during discussion about critical issues during his presidency. They choose a scenario, such as the strategy for Lincoln's reelection in 1864, and then utilize touch screen technology to explore documents that can help them decide how they would advise Lincoln to proceed. Visitors are able to direct their own experience as they navigate a series of choices about which issues to tackle, which character to become, and which sources and documents to consult.

Another creative use of technology at the visitors' center was the "Lincoln Yourself" experience. Visitors were invited to create a scene using their own face and making choices about what to include. When finished, the visitor received a description about the characteristics they chose. The completed image could be emailed to family and friends. In addition to providing a unique way to share information about Lincoln's defining physical characteristics, his surroundings, and items that helped define his interests and character, this museum feature provided an opportunity for visitors to share their experience at Lincoln's Cottage with others who were not with them and create a memento of the visit. Though the temporary exhibit that included this interactive feature has since been closed, visitors to the Lincoln Cottage's website can still, as of this writing, participate in the activity.[8]

The Place and Power of Museums in Washington

The challenge of presenting material that engages, educates, and preserves is hardly unique to Washington, D.C., museums. What is unique, however, is how many of them must also take into account the symbolic and political importance of their location, as well as the fact that the federal government has significant governing and budgetary power over a good many of them.

This burden is especially great for museums located directly on the National Mall and within the "monumental core" of the city, a highly symbolic place. Anchored by the Capitol Building and the Lincoln Memorial, the pedestrian friendly National Mall, sometimes referred to as "the nation's playground," has been seen as central to the city ever since L'Enfant designed the space as a major boulevard.[9] Today, the Mall is flanked on both sides by an impressive array of museums of art, history, technology, and ethnicity. The symbolic location of these museums—not to mention the status of the Mall as "the" place to visit a museum or gallery—puts a considerable onus on them to provide the best experience to visitors and represent their mission as fully as possible. It also makes the Mall extraordinarily desirable for other would-be museums seeking a prime spot in the nation's capital, though the National Capital Planning Commission—the body responsible for the development of the Mall—strongly encourages museum builders to look elsewhere in the city.[10]

The federal government has long had a role in the founding and governance of museums in Washington, D.C. The Smithsonian Institution, which runs many of the capital's museums (including some of the Mall's most famous facilities, including the Air and Space Museum and the Museum of Natural History), started with a bequest to the U.S. government from James Smithson, a British scientist who had never even been to America. His only stipulation was that the money be used to establish an institution for "the increase and diffusion of knowledge." Congress was closely involved in the initial decisions about how the money should be spent and the design of a trust to support the new institution. Congress never fully surrendered that authority: today's Smithsonian Institution is both a federal organization and a private foundation, run by a Board of Regents made up of private citizens and members of Congress, and approximately two-thirds of its budget comes from an annual Congressional appropriation.[11] Such authority extends to other museums in the capital too. Though not part of the Smithsonian system, the National Gallery of Art was also created by Congress, and the federal government has governing powers over the U.S. Holocaust Memorial Museum.[12]

This power has occasionally translated into a desire by Congress to get involved when museums display controversial contents. In December 2010, for instance, the National Portrait Gallery (a Smithsonian museum) removed a video art display by David Wojnarowicz when a top Catholic leader and a House Republican leadership aide both complained about an image of ants crawling over a crucifix.[13] An even bigger tumult involving Congress, discussed further below, took place in the early 1990s when the Air and Space Museum used a controversial display to feature the *Enola Gay* bomber from World War II.

Preserving National Identity

All museums and public displays provide an opportunity to share a story with the public. In Washington, many of the city's museums claim to be "national"—the National Museum of American History, National Gallery of Art, National Building Museum, and so on—suggesting that their content tells a story of the country or otherwise reflects national identity. More broadly, these museums have an additional, symbolic mission: to define what it means to be an American.

The museums on the National Mall, built at different times for different purposes, are a valuable historical legacy of how the nation has perceived this mission over time. The Smithsonian Castle, completed in 1855, was to house not only galleries but also science labs and other research and educational facilities (see figure 3.1). But the first building dedicated solely to serve as a museum—and originally called the U.S.

Figure 3.1 An Early Exhibit at the Smithsonian Building (1867)

Source: Prints and Photographs Division, Library of Congress, LC-USZ62-60427

National Museum—was constructed next door. The Arts and Industries Museum, as it was later named, was completed in 1881 and provided space to house and display objects from the 1876 Centennial Exhibition. It fulfilled the nineteenth-century mission of a museum to display objects of value—and in this case, objects that had come from an event celebrating the country's birth. The exhibits were designed to display objects as well as to educate visitors about the prevailing power of democracy. Early exhibits included a history of human civilizations, a geology and natural history exhibit, a hall of the history of technology, and a collection of personal items that belonged to Founding Fathers, called the Historical Relics exhibit.[14]

Facilities that followed, such as the Museums of American History and Natural History, were more specialized in content but similar in mission: to demonstrate the artifacts and items that symbolize a consensual American experience and distinguish the country from others. But the focus and content of the city's museums, similar to those throughout the country, were challenged by the social revolution of the 1960s and 1970s. As the country's image of itself changed, the content of museums followed. Many that had long adopted the perspective of the elite were forced to reevaluate their mission and message to incorporate ideas of multiculturalism, diversity, and the experiences of underrepresented and everyday people into their content and narrative. Museums incorporated new stories and perspectives to their existing collections and exhibits. For example, the National Museum of American History's exhibit A Nation of Nations, which was on display from 1976 to 1991, was specifically designed to "construct a multicultural image of America," celebrating "diversity as an enduring and defining characteristic of national life." It included a variety of artifacts telling the story of multiple immigrant groups and how they experienced life in the American nation. One section of the exhibit was designed to explore ways that immigrants from all over the world shared the "American" experience. Some of the objects on display included an entire public school classroom, a large flag made up of campaign pins from presidential elections since the Civil War, and a collection of neon restaurant and food signs. Each of these displays was connected to a broader theme about American life, including free public education, citizenship, and the way that the food industry reflects the diversity of the American people. The Smithsonian Institution also opened the Anacostia Neighborhood Museum in 1967 as an attempt to reach out to the local African American community in Washington, D.C., and minority visitors from other countries, in a way that the major museums along the Mall had not (see textbox 3).[15]

TEXTBOX 3: The Anacostia Community Museum

The Smithsonian Institution's Anacostia Community Museum originally opened in 1967 as the Anacostia Neighborhood Museum. This museum was designed to reach out to the African American community in the Anacostia neighborhood, hoping to provide an opportunity to preserve the heritage of the community while also connecting to the network of Smithsonian museums located in the city center. Smithsonian Secretary S. Dillon Ripley described this venture as an "experimental store front museum" located within the neighborhood it hoped to serve. Community leaders were involved in the development of the content and exhibits of the museum, as well as in raising support for the venture. An early exhibit focused on local rodent infestation problems, while other exhibits told the history of the Anacostia community and significant individuals from the area. The museum was successful: its innovative approach to local history, engaging in dialogue with the local community in the exhibitions, and providing a unique experience made the museum a prototype that was replicated around the nation, sparking the "neighborhood museum" movement.[16]

From the beginning, the focus of the Anacostia Museum was to build community identity and preserve that identity through its collections and exhibits. Though there have been numerous changes since its opening, including several name changes and even a new building, the Anacostia Community Museum has worked hard to focus on its mission. It still engages in community-based research and encourages local involvement in exhibit planning and design. Some of the subjects addressed in the museum go beyond a local story, but the museum's primary focus remains the people and neighborhood of Anacostia.[17]

Other new museums that explicitly aim to tell a broader narrative of the American experience have been proposed for the city. Two such museums—the National Museum of the American Indian, built in 2004, and the National Museum of African American History and Culture, currently under construction—demonstrate this shift quite starkly. Both fit within a national trend of "ethnic museums," focusing their collections, exhibits, and public programming around a specific group of people.

The theme of the National Museum of the American Indian (NMAI), part of the Smithsonian Institution, is "survivance": the resilience of tribes against tremendous challenges and adversities. The museum features several unique and distinguishing features as it attempts to incorporate both history and living culture in its exhibits. For example, the shape of the building is unusual: rather than follow the more traditional rectangular design of other structures along the National Mall, the building is curvilinear, celebrating the sacred form of the circle as understood by native peoples (see figure 3.2).

Figure 3.2 The National Museum of the American Indian

Source: Library of Congress, Prints and Photographs Division, Photograph by Carol M. Highsmith, LC-DIG-highsm-12698

Native Americans from numerous tribes throughout the country were also invited to participate in the process of designing the space and content of the three permanent exhibits in the museum. Told through this native voice, the exhibits do not follow a chronological order but are grouped by theme and, in some parts of the museum, by tribe. Artifacts are often displayed in an artistic way, with a minimum of explanatory text, so as to maximize their visual impact. Native peoples are also given the opportunity to tell their stories and share their experiences. There are video displays that feature recordings of members of select tribes discussing their beliefs, heritage, and way of life, and in some areas the screens have been used creatively to simulate interaction between the people on the screen and the visitor.[18]

The museum's bold structure and the novel organization of its material represents the diversity of perspectives within the American Indian community and forces visitors to rethink the purpose and meaning of a museum. Its nonlinear approach in particular leaves visitors free to dictate the order of their own experience in the spaces. Nonetheless, it is not without drawbacks. Its nontraditional exhibit designs can be challenging for visitors who, seeing a display of arrowheads or traditional baskets, are often given little information about their origin or meaning.[19] Some potentially controversial themes, such as missionary work among natives and the forced relocation of tribes, are only addressed in vague, symbolic ways without a clear narrative that could tell a different story than what is often given in textbooks. The biggest challenge for the NMAI, however, is its identity. A cultural institution must have a clear picture of what it wants to be and who its target audience is. Alas, the NMAI's goals are diverse, and in an attempt both to serve as a visitor destination and to provide a voice to native peoples, it has created a space that may be unable to fully accomplish both.

The other "multicultural" Smithsonian on the Mall, scheduled to open in 2015, is the National Museum of African American History and Culture (NMAAHC). The museum, the fruition of a campaign that began over a century ago to create a national African American museum, will be located northeast of the Washington Monument and will be devoted to African American history, life, art, and culture. In order to tackle complex issues of slavery, racism, migration, culture, and civil rights, the museum plans to confront the reality of racial oppression while also highlighting achievements of both famous and ordinary African Americans to "humanize" their story and reintroduce it into the national narrative.[20]

Though the museum has not (as of this writing) been completed, its plans indicate some similarities in content and design with the National Museum of the American Indian. In particular, the curatorial staff of the NMAAHC is working to involve those who are the subject of the museum—ordinary

African Americans—in the development of content. Its Memory Book Project invites blacks to share memories and images in a virtual scrapbook. In addition, it will work with an ongoing oral history project known as StoryCorps to record interviews with African Americans and store them at the museum and the Library of Congress. By providing platforms for certain peoples to record their stories and memories, the museum underscores its mission as a museum of *people,* adds a personal dimension to the space, and captures valuable stories that might otherwise be lost.[21]

As one might expect, other groups have come forward with proposals of their own for demographically or culturally specific museums on or near the National Mall, including a National Museum of the American Latino and a National Women's History Museum. They argue, not without merit, that they represent equally important groups of Americans who, like African Americans and Native Americans, have been historically disadvantaged and are ignored or underrepresented in other museums in Washington. By placing such museums on the Mall, they would also provide symbolic as well as substantive recognition of these people's contributions to the nation.[22]

But, leaving aside the practical issue of space—there simply isn't enough room on the Mall for all groups that might deserve a museum—these new initiatives demonstrate the potential dangers of culturally specific museums. True, they add diversity and complexity to the story of the nation. But by creating separate museums for individual groups, the idea of a unified American identity may be lost or weakened. As ridiculously limited as it may now seem that the Mall was once home to a single U.S. National Museum, such an institution at least recognized the value of a cohesive national narrative.

Constructing Memory

Beyond telling stories that preserve and enhance national identity, museums help keep memories alive. But in many ways they are also social agents that not only preserve history but also call visitors to debate, discussion, and action.[23] If a museum is the first place where an individual confronts a subject, it may help shape and preserve her memories related to that subject—yet also provide a more complex understanding of material by challenging her existing recollections and preconceived ideas. For many museums in Washington, D.C., the decision of which memories to tell, and how if at all to challenge existing understandings of the past, is magnified because of their "national" mission, their unique connections to the federal government, and their symbolic significance. It is further complicated by the fact that history is frequently more complex, nuanced, and morally ambiguous than people realize.

Take, for instance, the permanent exhibition space at President Lincoln's Cottage at the Soldiers' Home, located in upper northwest Washington. Given that it is the site where Lincoln is said to have written the Emancipation Proclamation, the museum gives significant attention to how the proclamation came about. But many do not understand Lincoln's complex and changing thoughts about race, slavery, and emancipation. Accordingly, the curators of the exhibit include a timeline of events and excerpts from different speeches, legislation, and writings by Lincoln to help visitors understand his thoughts on such difficult issues and how they changed before and during the Civil War.[24] By challenging visitors to consider the complicated history of slavery in the United States and how many Americans, including Lincoln, altered their thinking on an issue over time, those seeing the exhibit are encouraged to consider multiple perspectives, perhaps even sparking additional dialogue during and after the visit to the museum.

Challenging an existing public memory is not without peril. Perhaps the most famous instance of controversy surrounding the issue of museums and memory in recent Washington history was the so-called Battle of the *Enola Gay* in the mid-1990s. When the National Air and Space Museum was given the opportunity to display the fuselage of the iconic bomber *Enola Gay,* which dropped the atomic bomb on Hiroshima, the museum jumped at the chance. But in developing the narrative that would accompany the artifact, the museum's curatorial team and historians highlighted the horrific effect of the nuclear explosion on the people of Hiroshima and explored the question of whether the bomb necessarily ended the war, thereby saving the lives of American soldiers. American veterans groups were furious at this challenge to a historical narrative that emphasized Japan's wartime atrocities, the fierce fighting tactics by Japan which had led policy-makers to consider using a nuclear device, and the bomb's role in ending the war. Timed to coincide with the fiftieth anniversary of the bombing, many believed that the exhibit should be commemorative, not interpretative. After fierce criticism from the press, the public, and in particular members of Congress, museum director Martin Harwit resigned. To the chagrin of many historians, the NASM ultimately ended up cancelling the exhibit and replacing it with a new, simplified one that had the plane but little else. Ultimately, the result was an exhibit with which no one was happy.[25]

Other museums have managed to navigate the shoals of controversy when addressing contentious issues of history. When the United States Holocaust Memorial Museum, which opened in 1993, was first being planned, museum officials faced a daunting challenge. Created by Congress and slated to be built near the symbolically important National Mall, the museum's content—what to include and how to interpret events—would likely impact how people in America would remember the Holocaust. Tough questions had to be answered. Who would be involved in the process of creating the "preferred narrative," whatever that might be? What specific details would be included in exhibits, and should the harshness of events be toned down? Was the museum's space commemorative, interpretive, educational, or some combination thereof? And who was the intended audience: Americans or Europeans, Holocaust survivors or the broader public?[26]

As difficult as these questions were, museum curators and designers managed to find answers that have yielded one of the most praised and visited museums in the city. It draws visitors through a narrative that builds in intensity as they travel through the permanent Holocaust exhibit, with many elements (such as the ID cards given out at the start) designed to humanize events and tell individual stories amid an overwhelmingly destructive human catastrophe. Much of the museum's collection consists not of beautiful or valuable items that one might typically see at museums, but of ordinary, everyday objects that were never intended to survive very long—items such as shoes, clothing, and personal jewelry. What makes these items so powerful is that they have outlived the people who owned and used them. Finally, to connect an event that occurred in Europe to an American audience, museum designers highlighted American ideals including tolerance, equality, and freedom to create a sharp contrast with the Holocaust. One example can be seen in the exterior design, where quotations from speeches given by leaders have been engraved, evoking American values right from the beginning. There is also a quotation from the Declaration of Independence prominently displayed at the main entrance. At the same time, curators did not whitewash the less savory aspects of American history before and during the Holocaust. For instance, one display discusses the sad plight of the ship *St. Louis*, which left Germany in 1939 with nearly a thousand Jewish refugees. Despite the pleas of its passengers, the United States refused to accept them, and most were forced to return to Europe, where many died in the Holocaust.

Looking Forward

Museums in Washington, D.C., similar to museums everywhere, must confront the various needs and expectations of visitors while fighting to remain relevant in a changing culture. In addition, Washington museums are faced with unique challenges associated with the tensions involved in trying to simultaneously address local, national, and international audiences and the rising costs associated with providing quality experiences in an environment where visitors expect no or low admission fees at museums. Not all of them have been able to meet these challenges, and some have been forced to think creatively to survive.

The City Museum of Washington, D.C., which opened in May 2003, was one casualty of such challenges. Its purpose was to tell the story of the city of Washington, D.C., and it struggled from its inception to draw visitors, despite having secured a historic location and receiving financial support from private donors, the local government, and Congress, as well as the strong support of the Historical Society of Washington, D.C.[27] While its distance from other museums, admission fees, and a lack of advertising all contributed to the museum's failure, the biggest problem was really one of identity. The museum struggled to define the story it wanted to tell of the city of Washington and whether its primary target audience was local residents or tourists. As a result, the number of visitors was far less than had been projected, and the venture was not able to sustain itself. The City Museum was officially closed in April 2005.[28]

Financial challenges have impacted all museums in the city, even those within the vast Smithsonian Institution network. In order to subsidize the rising costs associated with creating large-scale exhibits in an era of decreased congressional appropriations for museums, Smithsonian officials have been forced to attempt creative collaborations and partnerships. One area that has been relatively successful, though not without criticism, has been corporate sponsorship. Visitors can see the effect of these financial partnerships throughout Smithsonian museums in Washington. Corporate donors often provide financial support for exhibits designed and marketed by the museums and receive recognition on signs or promotional material connected to the exhibit itself. For example, the America on the Move exhibit at the National Museum of American History received sponsorship from corporations such as the Automobile Association of America (AAA), General Motors, and the Association of American Railroads. In some cases, sponsors are involved in planning and executing an exhibit too. The first time that Smithsonian staff allowed this kind of participation was the Ocean Planet exhibit at the National Museum of Natural History in the mid-1990s. More recently, sponsors have supported special events, as well as underwritten existing exhibits. For example, in the spring and summer of 2011, children were invited to go on an American Girl tour through the National Museum of American History. Following the character Addy from the American Girl doll and book collection, children searched for objects and photographs throughout the museum to follow her story. Many criticize such sponsorship, worried that it allows corporate sponsors too much control over the content of exhibits. The Smithsonian has defended the practice, however, insisting that it follows firm and transparent guidelines for accepting and overseeing relationships with sponsors.[29]

Other attempts to provide educational experiences while reducing the financial burden can be seen throughout the city. Visitor centers and smaller-scale museum-style exhibits connected to, or built inside of, tourist destinations create opportunities to educate visitors without the costs involved in operating large museums. For example, the Library of Congress and the National Archives include both permanent and rotating exhibit spaces. The U.S. Capitol Visitor Center, opened in December 2008, has a permanent interactive exhibit telling the history of the Capitol and Congress, intended to provide context for visitors touring the Capitol building.[30] An extensive network of heritage trails and poster-sized street signs with historic photographs, maps, and narratives are placed around the city to offer visitors and residents opportunities to learn about local and national cultural history in individual neighborhoods.[31] There are also plans to build a Vietnam Veterans Memorial Visitor Center underground on the National Mall. Authorized by Congress in 2003 and officially approved in 2006, the space is intended to provide more history and context for visitors viewing the Vietnam Veterans Memorial Wall.[32]

Despite the many challenges facing museums and educational centers in Washington, the city's reputation as a "city of museums" seems secure. The fact that new museums are continually being built or proposed for the city underscores how important they are to the city's culture and identity. Our nation's capital, it seems, is destined to remain a center for preserving our nation's heritage for the foreseeable future.

PART II

Washington as Political City

Introduction to Part II

Politics exists in all cities—but perhaps none to such a degree, and on so many levels, as Washington. In the next four chapters, we consider the various political dimensions of the national capital. Two central themes emerge from this way of thinking about the city: first, that the capital offers a unique vantage point for observing national politics and political actors at work; and second, that national and international politics can have a profound impact on city life—sometimes, but not necessarily always, to the benefit of the city and its inhabitants.

National politics is the topic of chapter 4. In this chapter we consider the many ways that national politics manifests itself in Washington. This includes the city's reputation, how lawmakers and the president present themselves, and how members of the political community socialize and organize themselves. We also note how national politicians can make life difficult for Washingtonians—whether it be a presidential motorcade stymieing traffic or Congress insisting (until recently) that the city use a confusing taxi system.

Chapter 5 is about political protest. Unsurprisingly, Washington today is where groups and leaders of all stripes go to demonstrate on behalf of a diversity of causes. But for the first century of our history, public gatherings to press for political action were seen as inappropriate at best, a dangerous turn to mob rule at worst. A close study of the history of key demonstrations in the capital, both attempted and actual, reveals how and why those attitudes changed.

Washington is a place not only of national politics, but also of international politics. We explore this topic in chapter 6. The importance of prestige in international affairs explains the location and appearance of many embassies in the city. Being home to so many embassies also brings both unique benefits to Washington—culture, diversity, and economic investment—and unique disadvantages, ranging from the inability of residents to regulate the construction of embassy buildings to conflicts that can sometimes turn violent.

Finally, in chapter 7 we discuss local politics. Many elements of city politics are the same as in other American urban places: conflicts over limited resources, intense competition to attract wealth, and a focus on providing concrete goods and services to constituents. But Washington's legacy as a major African American hub, its constitutionally imposed boundaries, and Congress's considerable power over the city help explain what makes Washington's local politics truly unique.

CHAPTER 4

Institutions, Power, and Political Community in Washington

National politics permeates Washington. It can be seen and felt in many ways. Perhaps the most obvious way it manifests itself is in the city's reputation: Americans associate Washington with political power to such an extent that its very name is often used as short-hand to describe the entire national government. As a consequence, the capital's image, and its ability to draw or repel outsiders, has long been affected by citizens' views of political affairs and power.

But national politics affects Washington in other, less well-known ways as well. Ask any resident whose downtown commute has been blocked by the presidential motorcade: the behavior of those in power affects the life of everyday Washingtonians. In fact, such behavior—including the disruptive, multi-block-long White House motorcade—is often a visual manifestation of basic principles of American government. National politics also shapes the city of Washington through the practices, mores, and behavior of Washington's political community. Though that community's subculture is occasionally documented by journalists (and often negatively, such as in Mark Leibovich's book *This Town*), it remains a little-seen but no less important way that national politics shapes the city.

The Political City, Loved and Hated

Washington, D.C. is a far more economically diverse city than most realize; the vast majority of its residents work in the private sector, and many are in jobs that are, at best, tangentially related to the functions of government.[1] Nonetheless, to both city outsiders and the capital's political class, Washington has a single identity: a government town. As we noted in chapter 1, its public architecture—bold, dramatic buildings and monuments that seem to express power and authority—reinforce this identity. So too do smaller and more subtle objects, works of art, and images that can be found in the city. For instance, when a series of cow sculptures were displayed throughout Chicago in the late 1990s, Washington followed suit with donkeys and elephants, the mascots of the two major political parties. Advertisers also sometimes employ imagery and text that encourage Washingtonians to think of their city as a place of politics (see figure 4.1).

Because Washington is perceived as a political city, its reputation tends to be entwined with how Americans view their government. And that view is usually negative. America has a long cultural tradition of distrust toward political power and centralized authority.[2] That distrust was so strong and widespread in the early 1800s—a distrust that historian James Young called an "antipower ethic"—that even members of Congress openly complained about their work and the city of Washington and stayed in office as briefly as possible. In the 1830s, Alexis de Tocqueville validated the connection between Americans' skepticism toward government and the condition of the capital, assuring his readers that American democracy was secure because the country had "not yet any great capital" that might unduly

Figure 4.1 Politically Themed Advertisement in the Washington Metro

Photo by Matthew Green

exercise power. Americans even avoided using the word *capital* to describe Washington City until after the Civil War, fearing the word's nationalist overtones.[3]

Distrust of power remains a recurring element of American culture, as does the belief that Washington (and cities in general) are unseemly and possibly corrupt places. When hostility to government reemerged in the 1970s and 1980s, so too did a new pejorative phrase to describe Washington: "inside the Beltway," a reference to the 1960s-era freeway that encircles the city, which implies a certain distance from, if not a lack of connection with, ordinary citizens.[4] The language of contemporary representatives and senators reinforces and underscores that dislike of Washington and centralized power. For instance, lawmakers often run for reelection by campaigning against Congress or the national government (or simply "Washington"), even though they are technically federal employees themselves.[5] An incumbent may even be risking reelection if he settles in the city. In 2004, for example, Senate Majority Leader Tom Daschle of South Dakota narrowly lost reelection to a challenger who ran a campaign ad that featured Daschle declaring, "I'm a D.C. resident." Some lawmakers sleep on folding cots in their own offices rather than rent or buy property in the district, a way of declaring to constituents that they will not be "captured" by the city. One congressman, Jason Chaffetz of Utah, even records periodic video commentaries on politics from his office bed, calling them "Cot-Side Chats."[6]

Paradoxically, while Americans may distrust their government and the city in which it resides, they also find Washington a highly attractive place to visit. A trip to the capital is akin to making a patriotic pilgrimage to the heart of democracy, and for many it is an affirmative act: a way of finding the core of, and thereby expressing loyalty to, their country.[7] For some, it may even come from a desire to be close to power. This need to see, if not possess, power—a need as old as humankind—creates a gravitational pull that not only draws tourists but also new, youthful, and ambitious residents to the area. The novelist Henry Adams captured this well in his 1880 political satire of Washington, *Democracy*. As the book begins, its widowed heroine, Madeline Lee, seeks to live in the city because

> she wanted to see with her eyes the action of primary forces; to touch with her own hand the massive machinery of society; to measure with her own mind the capacity of the motive power. She was bent getting to the heart of the great American mystery of democracy and government . . . What she wanted, was *power*.[8]

Congress and the Presidency on Their Home Turf

So what does the city of Washington teach us about this "great American mystery" that both attracts and repels? Madeline Lee and another character from *Democracy,* her sister Sybil, quickly discover that the oratory of national politicians reveals very little about the politics of American government.[9] One could take a tour of the Capitol instead, or perhaps visit the White House, but that is unlikely to be terribly enlightening either. One can, however, get a glimpse into the political logic of Congress and the presidency by observing how their occupants present themselves in, and sometimes even try to influence, the city of Washington.

Take first the U.S. Congress, America's supreme legislative body. It is where representatives gather to draft laws, oversee the executive branch, and help their constituents navigate the federal bureaucracy. Congress is also a highly parochial institution: its members are elected from discrete districts or states, and they owe their jobs to the voters who brought them into office. Since most lawmakers want to keep those jobs, a great deal of congressional politics is structured around securing the loyalty of constituents: voting the right way on bills, helping constituents navigate the government bureaucracy, and getting federal funds for local communities (known as earmarks or, more pejoratively, "pork").[10] One of the best ways to observe this parochialism is to visit a representative's or senator's D.C. office. More often than not, the main lobby will be decorated with a map of her district or state, pictures of local landmarks, and objects the area is known for, be it wine, machine parts, canned goods, or sports memorabilia.[11] These items show off the constituency—and perhaps, similar to the décor of a travel agency or visitor's bureau, they encourage outsiders to learn more about, or even consider a visit to, the area. But more importantly, they transform the space into a small pocket of the state or district, which proclaims the lawmaker's allegiance to the area she has been chosen to represent. The result is a tribute to the legislator's voters and a way to proclaim her "likeness" with constituents.[12]

Another, less savory aspect of congressional politics—lawmakers' use of power for personal gain—can be measured by how it has made its mark on Washington affairs. One example is the city's informal practice of allowing legislators to avoid arrest for various crimes such as driving under the influence.[13] A more concrete example is the city's mammoth Rock Creek Park, which winds northerly through the western part of the District of Columbia. The park was created by an 1890 act of Congress with the help of Sen. William Stewart (R-NV). Senator Stewart, as it happened, had a financial stake in the company that was developing nearby neighborhoods and thus would profit if the park made the surrounding area attractive to home buyers. Stewart was hardly the only lawmaker who made personal real estate investments in the city in the late nineteenth century, despite the conflict of interest that resulted from Congress having exclusive jurisdiction over the District of Columbia.[14]

Legislator self-interest even extends to the city's taxi cabs. For decades Washington's taxis ran on a "zone" system in which fares within each zone were constant. The city's large central zone happened to include the Capitol, all congressional office buildings, the main train station, the White House, and even the trendy neighborhood of Dupont Circle (see figure 4.2). It was a convenient arrangement for lawmakers and presidential staff, who paid a single low fare to visit a wide range of important locations. Congress routinely passed district funding bills that prohibited any replacement of the scheme with a traditional meter system. The zones vanished only after a U.S. senator who happened to dislike them, Carl Levin of Michigan, added language to a spending bill requiring meters in D.C. cabs.[15] To many Washingtonians, it

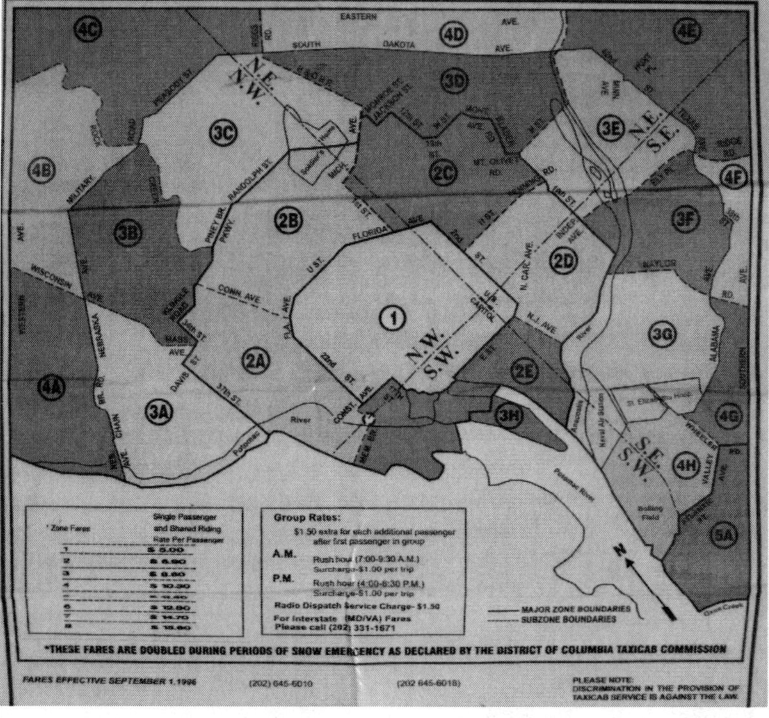

Figure 4.2 Old Washington Taxi Zone Map

Photo by Matthew Green

often seems that members of Congress are like Senator Levin: caring about the district only when something about it bothers them personally.[16]

In contrast to the parochial politics of Congress, the politics of the presidency are largely about symbolism. The president has far less formal authority than most people realize. The Constitution gives him a few key powers, including the legislative veto, delivery of the State of the Union address, and the right to "recommend" actions to Congress. But presidents must share other powers with Congress (such as appointing federal judges) and are denied some powers that executives in other countries have, such as the line-item veto (the right to selectively veto parts of bills) or calling for new elections. Furthermore, a vast amount of the White House's time is spent reactively, putting out fires or conducting political triage after a sudden crisis occurs, creating a "sense of impotence" in the West Wing. Presidents' primary leverage comes instead from the *perception* of power. "Presidential power is the power to persuade," as White House scholar Richard Neustadt put it, and projecting an aura of authority—establishing strong public prestige and making strategic public appearances to push Congress in a desired direction—is crucial to that power. Congress tends to defer to newly elected presidents because of the perception that they have a "mandate" from voters, and even strong opponents of presidents speak of the intimidation and sense of awe they feel when the president calls or meets them personally.[17]

Symbolism matters to the perception of power. How presidents are seen in the public sphere, especially when in motion, serves to represent and thereby heighten an impression of political authority. When the chief executive moves around the country or the world, advance teams are employed to secure his next location; he travels by personal jet or helicopter with a massive security entourage; and international communication links and a large coterie of staff go with him. The dramatic physical presence of this mobile security apparatus is visible in Washington, too. City traffic is routinely delayed or blocked by the presidential motorcade, which includes police motorcycles, Secret Service cars, and even an ambulance. His Marine One helicopter flies overhead as the president travels to and from the District of Columbia. Even a private visit to a local restaurant by the First Family requires dozens of Secret Service agents to come in advance and scan other restaurant guests with portable metal detectors.[18]

This is not to say that presidential security is purely for show. The point is that, besides demonstrating the level of personal threat that the president faces, these actions signify the tremendous value that the nation puts on the presidential office and—intentionally or not—heightens the aura of power that surrounds it. Even before the president's safety became a major concern, presidents conveyed their importance by travelling with many personal aides. The columnist H. L. Mencken recalled that "even so stingy a fellow as" President Calvin Coolidge "had to hire two whole Pullman cars to carry his entourage" of staffers, reporters, and security personnel. The president's moving security team is, in other words, an example in Washington of what the anthropologist Clifford Geertz called "symbolics of power": one of the Oval Office's many "stories, ceremonies, insignia, formalities, and appurtenances" that are needed by the rulers of any organized society to convey their authority.[19]

What Is Washington's Political Community?

Congressmen, senators, and the president are only some of the individuals who together make up the city's community of national political actors. Others include cabinet officials, policy advisors, legislative aides, political consultants, reporters, interest group leaders, and lobbyists. What can we say about what Leibovich calls "This Town"—the larger group of people who live, work, and socialize in the greater Washington area?

The two classic images of a Washington politico are as follows: (a) an older, white male politician or politician-turned-lobbyist, and (b) a single, recent college graduate who works in politics for a few years before starting a long-term career elsewhere. These two stereotypes do roughly match one subset of Washington's political community: those elected to, or who work in, the House or Senate. Among Congress's elected members, over 80 percent are male and over 80 percent are white. The average congressional staffer may be male or female but—at least in the House of Representatives—is often young and usually has a college degree, and only about 40 percent of House staffers are married (versus 60 percent of workers nationwide). The average tenure of a legislative aide in Congress is five to six years, with 60 percent having worked less than two years in their particular job.[20] The significant churn in congressional staff, and the large numbers of youthful legislative aides, interns, and other political workers who come from outside the city to work on Capitol Hill, has contributed to an image of Washington as a "transient" city, a reputation first earned in the 1800s when its population was dominated by lawmakers who were frequently defeated for reelection.[21]

But political Washingtonians are not all transients or congressional employees. There are roughly 15,000 staffers on Capitol Hill—a big number, to be sure—but the House and Senate have only 535 voting members, and there are some 12,000 registered lobbyists in Washington, over 2,000 journalists accredited to cover the U.S. Senate or the White House, and nearly 1,000 people appointed to top cabinet positions or working in the White House.[22] Nor is Washington any more likely to have new residents, or residents from other states, than cities such as Boston or San Francisco.[23] In fact, many members of the city's political community are older, married, and long-time inhabitants of the area who have worked in a variety of jobs, both public and private sector. Among the more famous individuals who fit this profile include Clark Clifford, a presidential legal aide who served four different presidents in varying capacities and later became an influential lobbyist, and Daniel Patrick Moynihan, who worked as an assistant secretary of labor for two presidents, an advisor for a third, an ambassador to India and the United Nations, and a senator for twenty-four years.

Such a large and (somewhat) diverse group defies easy generalization, apart from the predominance of white men in its membership.[24] Nonetheless, since Washington's earliest days the most prominent members of the city have followed certain sociological patterns in how they interact and express themselves in the broader community. These include establishing social groups, putting individuals into hierarchical rankings, forming strategic interpersonal connections, and socializing at designated events and gathering places to reinforce those groups, hierarchies, and connections. As we shall see, sometimes these patterns even influence the decisions of government itself.

Social Groups and Hierarchies in the Political Community

Humans are social animals, and politics is a social vocation. Unsurprisingly, then, members of the capital's political community have tended to form into social groups: structured sets of individuals who interact, have a communal feeling of belonging, and share particular interests or goals. In the city's first few decades, lawmakers tended to live near where they worked, forming distinct and separate communities held together primarily by common regional ties.[25] Today, people of similar profession, workplace, or political party—as well as place of origin—still tend to form into distinct groups. *Peer groups*, in which members are relatively equal in age and status, are also common in Washington.[26]

In addition to forming social groups, a community of reasonable size will usually establish internal hierarchies based on some given criteria. Washington's political community is no exception. Rankings have long formed within the city's political class based on what the anthropologist Ralph Linton called *achieved status*, or status based on profession, reputation, or other factors determined by "competition and individual effort." Those considered politically influential because of their jobs or connections are usually ranked higher than those considered less powerful.[27] The person at the top is the chief executive. "If . . . Washington is a company town," wrote former *Washington Post* owner Katharine Graham, "then the president is the head of the company.

Therefore, he has always been an important man in Washington, a big man around town." Representatives and senators are close behind: in a statement from over a century ago but no less true today, one observer noted that "an election to the House of Representatives gives to the fortunate man a *prima facie* right to respect and consideration" in Washington society.[28]

Hierarchies may be less important than they were in the nineteenth century, when the city's high society was central to the lives of so many political actors. But hierarchies never entirely disappeared, either. In the early 1940s, the wife of a leading Washington journalist hosted an informal picnic for many of the city's leading figures. No seats were formally assigned, but she later wrote, "To my astonishment I saw that every person had seated himself or herself exactly according to protocol." The hierarchical, status-oriented nature of the political community still exists and continues to be internalized and enforced by its members. So durable is the fixation on one's rank that more than one observer has compared Washington to high school, where students form pecking orders based on merited, arguably trivial criteria such as how long one has been at school or one's extracurricular activities.[29]

"People here are often defined by what they do," wrote Katharine Graham, and a social hierarchy requires methods to identify others' professions and advertise one's own. Small talk at social functions is frequently punctuated by the query, "So where do you work?"—the answer revealing one's likely political influence and peer group (and, for the snooty, whether one should even continue the conversation). Also common is what the sociologist Talcott Parsons termed an *expressive symbol*, an object used to communicate status or identity. The business card, ubiquitous in Washington, is one such signifier of profession and thus hierarchical ranking. Another is the lapel pin. The pin, originally offered as an option to U.S. senators, was required for all House members starting in 1975 to help Capitol police distinguish lawmakers from the growing numbers of congressional staff (and probably also to help them identify the dozens of freshmen lawmakers elected that year).[30] It is today worn by Washington lobbyists, businessmen, and others who, clothed in similar business attire, need to distinguish themselves and advertise their professional affiliation.

Social Connections and Events in the Political Community

As members of D.C.'s political community interact with others from various social groups, connections and contacts invariably follow, and a social network is formed. Connections and networks are something of a necessity for those in elected office, particularly members of Congress. American government, with its many participants and with powers divided and shared among different branches, depends on the "cooperation . . . of many other players" to work. Any lone politician who wants to go it alone is likely to become "isolated and feeble," as Washington journalist Hedrick Smith put it.[31]

Social networks were especially vital during the capital's early years, when political institutions were new and most of its residents were strangers. The historian James Sterling Young argued that the separate residential communities that emerged around the Capitol and White House in the early 1800s formed internal social networks that helped enforce the constitutional division between the legislative and executive branches. Several early arrivals to the city founded a nascent social scene that "was all about politics, in every sense of the word." Samuel Smith, a chronicler of congressional debates and later secretary of the treasury, and his wife Margaret Bayard Smith became one of the city's first "power couples," entertaining guests at their rural home (on what is now part of the campus of Catholic University) and, later, in the city. Perhaps the most important contributor to the growth of political society, however, was Dolley Madison, who arrived in Washington in 1801 as the wife of President Jefferson's secretary of state, James Madison. Charming and savvy, Dolley "actively drew people to her" with "generosity and openness." As one of the first and most important political hostesses in the city, she created an informal sphere of politics that made it easier for politicians from different parties and branches to build connections and work together and thereby "contributed to the construction of a workable government in Washington City."[32]

Though the rules of engagement for social events among Washington's elite have undergone many changes over time, they have usually reflected, if not helped reinforce, hierarchies within and boundaries around the political community. In the city's first decades, dinners and, more often, social calls—leaving one's calling card with another person's servant to announce one's presence—were a common means of socializing, and it was expected that a new resident's first social call would be to the home of someone higher in rank.[33] During the late nineteenth century, an influx of wealthy Americans built second homes in the city. Social life subsequently expanded to include formal balls and large receptions, giving rich Washingtonians an opportunity to demonstrate what economist Thorstein Veblen called the "conspicuous consumption" of the "leisure class." Select members of the political community, particularly its most renowned, either sponsored or were expected to attend such events. Wives of the economic elite became premiere entertainers. Evalyn Walsh McLean, daughter of a wealthy miner and wife to the man whose family owned the *Washington Post,* threw huge parties in the early 1900s at which she would sometimes show off her most famous possession, the Hope Diamond. Gwendolyn Cafritz, wife of a renowned real estate developer, was a preeminent social doyenne from the 1940s to the 1970s. When a letter appeared at Washington's main post office addressed only to "Capital's No. 1 Hostess," it was dutifully forwarded to her. Though African Americans were excluded entirely from this social scene until the mid-twentieth century, black Washingtonians developed separate hierarchies and social networks and hosted their own formal events.[34]

Formal private get-togethers are less central to today's Washington society than they once were, but they still happen. Top lobbyists Heather and Tony Podesta, newspaper columnist George Will, and ambassadors and the wives of lawmakers are among those who host dinner parties for the "who's who" set in Washington.[35] White House gatherings remain the most sought-after in the city, but others are also prominent. For many years, the senior news correspondent for ABC, Sam Donaldson, threw an annual Christmas party that was "a favorite of Washington's movers and shakers," in the words of one former senator. In addition to private events at one's home, certain Washington establishments are popular locales for community members to socialize and make connections. The most exclusive are private organizations including the University Club, the Metropolitan Club, and the Cosmos Club. But for those who are not members of such clubs, certain restaurants and bars are understood as places where members of particular peer groups regularly meet. Young conservatives, for instance, hold regular happy hours at Union Pub on Capitol Hill, while Democrats often patronize an establishment on U Street called Local 16.[36]

Perhaps the most important (or at least the best-known) social events attended by members of the political community are annual dinners sponsored by media organizations, including the Gridiron Club, the Washington Press Club, the Radio-Television Correspondents Association, and the White House Correspondents' Association. They are fancy, sometimes glamorous affairs. For example, the White House Correspondents' Association Dinner evolved from a somewhat low-key gathering to a Hollywood star-studded "nerd prom" made (in)famous by Steven Colbert's satirical 2006 keynote speech lampooning President George Bush. Less visible are the media companies that spend tens—even hundreds—of thousands of dollars to entertain guests at pre- and post-correspondents' dinner soirees, making it a veritable "schmooze-a-palooza." More than flashy parties, these dinners permit community members to relax and socialize with each other and allow the newest among them to establish contacts and make good first impressions. The Press Club Foundation dinner, for instance, is "the annual opportunity for newbie lawmakers to introduce themselves to D.C. via witty speeches," an important move since "new reps may be big fish in their districts, but most are guppies when they first come to Washington," wrote two *Washington Post* chroniclers of the city's social life.[37]

Interpersonal connections, and the events and meeting places that help fortify them, enforce individuals' sense of social identity and strengthen the sense of in-group membership for members of the political community.[38] Conventional wisdom, at least in Washington, is that these networks and social gatherings have significant effects on politics, too. As the wife of journalist Raymond Clapper put it in 1946:

> It is true that social life in Washington frequently influences the business of government—seldom in a sinister way, but in the manner of congenial friends meeting often who like to do favors for each other. Therefore it is usually important to entertain and be entertained by people in influential positions, if you want to succeed in business.[39]

There are numerous examples of this influence in Washington's history. President Jefferson hosted evening meals for politicians that were "the talk of Washington"; some lawmakers eagerly curried Jefferson's favor in the hopes of securing a rare invitation. The president exploited his dinners to lobby lawmakers and secured valuable information from them, thereby subtly leading Congress. Dolley Madison used her vast network in many powerful ways, especially when her husband became president in 1808. She tempered political differences that threatened to explode into major conflicts, made strategic social calls to lawmakers and cabinet officials to burnish her husband's reputation, lobbied successfully to keep the damaged capital in Washington after the War of 1812, and was central in deciding who received certain government jobs. Fast forward to the twentieth century: in the early 1980s, First Lady Nancy Reagan made a surprise appearance at a Gridiron Club dinner to try to dispel her image in the press as a "frivolous clotheshorse who hobnobbed with the idle, partygoing rich." Her unexpected, self-deprecating, and hilarious performance—a surprise song-and-dance number lampooning that image—astounded the reporters in the audience, and media coverage of Mrs. Reagan became far less critical.[40]

The connections that emerge from these social events, gatherings, and exclusive settings can also evolve into genuine friendships. In fact, having close friends is especially valuable for the psychological survival of those in the high-pressure, often insecure world of national politics. But self-interest, more often than not, drives the initial formation of connections in the political world. Ambitious Washingtonians are inclined to seek out contacts because they want a prominent political job, need a favor, or wish to improve their social rank. As President Harry Truman cynically remarked: "If you want a friend in Washington, get a dog."[41]

Conclusion

National politics influences Washington's identity and reputation, how the city's political actors behave, and the social life of its political community. That behavior and social life in particular has occasionally raised troubling questions about American democracy in practice. For one thing, the community's inward focus may move policy outcomes away from the will of voters. For decades after the city was founded, for instance, anti-slavery activists argued that the capital's southerly location made it too easy for southerners to influence the national government. Until the 1870s, the perceived geographic and moral distance of the city from the nation even stimulated efforts to move the capital from the District of Columbia altogether. The desire to relocate the government is gone, but not the sentiment: one editorialist recently opined that Congress failed to do enough to combat the 2008 recession because Washington's "governing elite" was barely scathed by the economic downturn.[42]

Though less likely today, the social scene among politicians in the past could also interfere with the work of government. When President Andrew Jackson's secretary of war, John Eaton, married a newly widowed and younger woman rumored to be of loose morals, the president's cabinet became deeply divided over the so-called Peggy Eaton Affair. Gossip threatened to consume Washington society, and Mrs. Eaton was ostracized by the wives of powerful officials, forcing Jackson to demand the resignation of his entire cabinet. In the 1870s, President Ulysses S. Grant's nominee for chief justice of the Supreme Court (his attorney general, George Williams) was tarnished not only by accusations of bribery and theft but also by a major social faux-pas committed by his wife: she had demanded that the spouses of senators pay her their first social call, rather than vice versa. One of the senators whose wife was especially insulted sat on the committee considering Williams' nomination; at one hearing he rose "at his wife's prodding" (or so it was rumored) and spoke vehemently against both Williams and his wife.[43]

The possibility of social soap operas overwhelming the work of government seems fanciful today. And defenders of political socializing argue that it has many benefits—and that there should be more of it, not less. For example, former *Washington Post* owner Katherine Graham, known for sponsoring private dinners for important city visitors, argued that socializing between the media and politicians "is constructive and useful for both sides," helping the press gather information while providing policy makers with contacts to whom they can later suggest ideas or offer complaints.[44] Nonetheless, concerns about the social behavior of Washington's political elite remain—especially their high-profile dinners with Hollywood stars. Former NBC news anchor Tom Brokaw, for instance, worried that the White House Correspondents' Dinner "separates the press from the people that they're supposed to serve, symbolically." Washington columnist Dana Milbank criticized the Press Club dinner, the city's preeminent political gatherings of the press and politicians, this way:

> Awash in lobbyist and corporate money, it is another display of Washington's excesses . . . the cumulative effect is icky . . . [giving] Americans the impression we have shed our professional detachment and are aspiring to be like the celebrities and power players we cover.[45]

Whether Graham, Brokaw, or Milbank is closer to the truth—whether the political community's social mores are healthy for American democracy or a source of undemocratic and elitist behavior—the long history of political socializing in Washington suggests it will be a part of the city and a factor in American politics for the foreseeable future.

CHAPTER 5

A Center of American Protest

Two blocks from the White House, at 1401 Pennsylvania Avenue N.W., stands the Willard InterContinental Hotel. Through its long history, which began in the early years of the nineteenth century, the hotel, known as "the Willard," has served as an informal center of political power while playing host to countless visitors.

These days, in the grand atrium of the elegant building, there is a small plaque that notes the hotel's claimed contribution to the culture of the nation's capital. The plaque honors the legend frequently told about the eighteenth president of the United States and the lobby of the Willard. Legend has it that President Ulysses S. Grant would often walk to the hotel to enjoy a cigar and a brandy, but constantly had to avoid the political wheelers and dealers who gathered in the lobby seeking his attention. Accordingly, Grant gave the attention seekers a name that has stuck ever since: "lobbyists."[1]

Though the word had been in use decades earlier, it is fitting that the term is believed to have been coined in the national capital. Washington, D.C. is full of lobbyists—thousands of people employed as professionals who make their living trying to get the government's attention on behalf of a particular group, individual, organization, or entity. But if we define *lobbying* more broadly to include any exercise of one's constitutional right to petition the government, there is another important and common method of lobbying the national government that happens in Washington: the public protest. The city plays host to scores of protests and demonstrations every year, many organized around a common goal of changing public policy. Whether they successfully achieve that goal or not, something about articulating a grievance or making a demand while standing on the Capitol lawn or in front of the White House seems to resonate with Americans.

The Constitution's protected right to "assemble" and "petition" might suggest that a city created to house the federal government was a fitting place for political protest from the very beginning. In fact, the public, policy makers, and the press did not see public demonstrations as acceptable political activity—in the capital or anywhere else—until the early twentieth century. Lucy Barber, a historian of protests in Washington, argues that public opinion about the city as a national public space evolved over time.[2]

In this chapter, we examine two major protests in the nation's capital and public perceptions of them, as revealed by news reports, to demonstrate the evolution of public attitudes concerning the use of the nation's capital for public protest. One of those protests—the 1963 March on Washington for Jobs and Freedom—intersected with another central feature of the city: its large and politically aware African American population.

Historical Background

Collective action has long been a part of American history. Danver, the editor of an encyclopedia about American protests, reminds us that protest on American soil began with the Pueblo Revolt against European settlers in 1680.[3] The American Revolution was marked by several protests and demonstrations against British rule. But the success of the American Revolution was not followed with tolerance for demonstrations in the nation's capital. When the U.S. Constitution was amended to mandate that "Congress shall make no law . . . abridging . . . the right of the people peaceably to assemble," such assembly was not expected to include protests in the District of Columbia. The Founding Fathers had been rattled by protests against the national government in 1783 by former soldiers and again in 1787 by Shays' Rebellion, an uprising by unhappy farmers in Massachusetts. In fact, one advantage of placing the nation's capital distant from existing cities was to avoid the "tumultuous populace" of existing urban centers, as noted in chapter 1.

Protests, occasionally resulting in attempted rebellions, did occur in nineteenth-century America, though rarely in Washington. In 1877, angry supporters of Samuel Tilden, the Democratic presidential nominee who lost following a controversial vote count, planned a protest outside the U.S. Capitol. The protest was canceled, however, when Tilden spoke against it and outgoing President Ulysses S. Grant threatened to deploy soldiers on the Capitol grounds. The following year, police did allow two men to speak against Chinese immigration on the Capitol steps, but their effort drew only a small number of spectators.[4]

Coxey's Army

The first large protest in the District of Columbia was organized by Jacob Sechler Coxey in 1894. The businessman from Ohio decided that a "petition in boots" was the appropriate response to a government he saw as unresponsive to high unemployment. Coxey's specific demand was to "end the suffering of unemployed workers by building modern roads . . . and funding new community facilities with federally subsidized bonds."[5]

Coxey's decision to march to Washington came at a critical time in the history of American journalism. In this instance, journalists wanted big stories for their newspapers and magazines. In the competitive environment of the late nineteenth century, coverage of the events dubbed "The Army of Peace" and "Coxey's Army" by the press "generated the most newspaper coverage of any event since the Civil War."[6] Reporters were paid based on the number of inches of copy that appeared in print, sometimes tempting them to create stories when there was no real news.[7] Additionally, the invention of the telegraph made it possible to cover a story that began in one section of the country and ended hundreds of miles away.

It was within this context that Coxey conceived his plan to march to Washington to draw attention to the unemployed. Coxey was acutely aware that most Americans saw unemployed men as "tramps" or "vagrants" subject to arrest in most parts of the country and were unsympathetic to their plight.[8] Coxey, who had made a fortune selling quarried sand to steel and glass furnaces, partnered with a person who knew how to generate publicity: Carl Browne. Browne's background as a former journalist and a seller of patent medicines, in addition to his unusual religious beliefs—he told reporters he was a reincarnation of the soul of his dead wife and part of the soul of Jesus—made him uniquely qualified to draw attention to an event of size and controversy. Coxey and Browne adopted a strategy of deliberately crafting a "symbiotic" bond between the press and the army that "was figuratively a deal with the devil."[9]

The two leaders began their courtship with the news media by issuing a pamphlet in January 1894, paid for by Coxey, that called for a mass meeting on May 1 on the steps of the Capitol in Washington, D.C. The pamphlet was printed in a Massillon, Ohio, newspaper and was introduced in Congress as a bill sponsored by a California member of the House of Representatives who represented Browne's home district and by a Populist senator, William Peffer of Kansas. Unsatisfied with the amount of attention that had been generated, Browne went before the Massillon city council and asked for a meeting of citizens to consider the proposal. At the mass meeting, he engineered the appearance of approval by asking for a show of hands of those who supported the effort. (As one newspaper reported, he was clever enough not to ask for a show of hands from those who did *not* support the idea.[10]) The first official newspaper mention of the "petition in boots" to the nation's capital showed that the march could garner attention. The paper printed a front-page story of Browne's attempt to organize a massive march on the city council that was thwarted by the council's decision to adjourn early.

In the weeks before the march, Coxey and Browne showed stacks of mail from around the country to reporters who arrived in a steady stream to cover the novel event. The two said they expected hundreds of thousands to join what they called the "Commonweal of Christ." Journalists expressed skepticism, and so when just 122 people headed out of Massillon on Easter Sunday, March 25, they were more than happy to point out the discrepancy and describe the participants in harsh terms.[11] After noting that the eight-mile trip from Massillon to Canton, Ohio, resulted in a loss of some twenty-five members due to a snowstorm, the *Washington Post* reported the "motley procession" was treated by onlookers as a joke. The dispatch described the scene in the following way:

> Carl Browne, chief marshal, who headed the procession, was mounted on a white horse, and was followed by half a dozen aides, all mounted on horses belonging to Coxey, who rode in carriage drawn by a pair of spirited steeds. The procession consisted of the marshals, Coxey, his wife and sister, a bugler, four covered wagons containing camping outfits, baled straw and several quarters of beef; a brass band that played all kinds of music at once, and the soldiers of the Commonweal on foot. They marched single file, or two abreast, as pleased their fancy, and, with very few exceptions, were hard looking citizens.[12]

The *Baltimore Sun's* story the day before the march began noted the high expense of this media coverage. The paper pointed out that the twenty-five or so reporters who made the trip were always accompanied by four telegraph operators and one lineman. Those resources added up to an expenditure of $6,000 for daily expenses for each person and $5,550 for the 1.1 million words telegraphed during the trip.

For the most part, Coxey and Browne got the news coverage they sought. But it came with a price. Though Browne consistently reminded the marchers of the importance of the press, marchers were not oblivious to the contempt the journalists felt for them. On several occasions, Browne had to intervene in physical altercations between the two groups. Even Browne succumbed to frustration, dubbing the forty or so reporters who accompanied the marchers "the argus-eyed demons of hell." It was a name they relished and used to mock the religion-focused organization and its marchers. The reporters promptly created demon uniforms, issued demon badges, and formed a "demon organization," which created marching orders that included scheduled times for whiskey and beer drinking.

Though Browne spoke out against the press, he also "enjoyed their company, respected the value of the news coverage, and once decorated the demons."[13] He also tried, with some success, to improve the press coverage. Reporters went along with the deception that a person called "the Great Unknown" was at the top of the march's leadership structure, although they knew it was Browne. During the march, photographs of members of the army taking baths in local streams, washing clothing, or getting their hair trimmed were published in newspapers to counteract the perception of the marchers as filthy "vagrants." Browne also permanently suspended three members of the army for allowing themselves to be put on display in a dime museum. Though the move overshadowed his and Coxey's attention-seeking plan of having an African American, Jasper Johnson Buchanan, bear the American flag at the start of the Washington, D.C., procession, the decision to "integrate the army raised eyebrows" and demonstrated "their news media savvy."[14]

When the Commonweal of Christ finally made it to Washington, D.C., on the first day of May, the media's criticism of the protest was severe. Many stories emphasized the fact that the march ended prematurely when Coxey, hoping to give a speech at the Capitol, was instead arrested for trespassing on the Capitol grounds. Under the headline "Climax of Folly," the *Washington Post* characterized the effort with the lines, "There was a great crowd, there was something of a procession, and there was the semblance of a riot."[15] A front-page story in the *Washington Times* emphasized the anticlimactic aspect of the event. The article began with "the coxey (sic) army came, saw, was clubbed and two of its leaders hustled off to the police station."[16] Coverage from the *New York Times* was even more critical and mocked the members of the procession. The front-page story began with the following sentence:

> Jacob S. Coxey's much-advertised demonstration on behalf of the "Commonweal of Christ," in favor of good roads and the repudiation of national obligations to pay interest on bonds, ended today in a fizzle quite as ridiculous as the principles enunciated by the leaders of the movement.

The story called the marchers "a disreputable crowd of tramps audaciously claiming to be the representatives of millions of respectable wage earners" who had "spruced up a bit for the great parade, but they were a sorry-looking set, with their broken shoes and ragged clothes." Not only did the story point out the huge gap between the number of marchers promised by Browne and the actual number who appeared ("six hundred of a promised 300,000"), but it made a point of dismissing one group of cheering bystanders in the city because "half" of them "were negroes."[17] In fact, the crowd that came to watch the event was forty times the size of the actual march.[18]

The main objections to Coxey's protest were rooted in concerns about public order and safety. Coxey spent the day before the march in Washington meeting with the district commissioners, and he made visits to the Capitol and the district buildings and the health office to obtain a permit for a parade.[19] According to one newspaper, his general reception was cordial, but officials were adamant in their opposition to Coxey's plan to address his "troops" from the Capitol steps.

Though ultimately Congress refused to consider the legislation proposed by Coxey and his supporters, the effort could hardly be considered a failure. Coxey's "unprecedented claim that ordinary Americans had a right to voice their demands in the capital" was the beginning of a change in how protests would be viewed in Washington.[20] The extent of the press coverage allowed what might have been a single event—the nearly four hundred-mile trek from Ohio to Washington, D.C.—to become a national issue that had political implications.

Protests in Washington after Coxey

Washington politicians and city officials learned from the Coxey affair, alternating between strict control and largesse toward future organized demonstrations. The next large protest in the capital, which came nineteen years after Coxey's Army, acknowledged the concerns of city officials by calling the event a pageant rather than a march. The Woman Suffrage Procession and Pageant took place in 1913, deliberately timed to occur the day before the inauguration of President-elect Woodrow Wilson. Though many bystanders and even some police mocked its participants, the otherwise-peaceful protest "emphasized how the beauty and dignity of the participants could carry over to public life in general."[21]

For city officials, the worst-case scenario involving political protest took place during the Veterans Bonus Army March of 1932, when 43,000 people came to the nation's capital in an effort to force Congress to pay World War I bonus money to veterans of the war. Concerned that the marchers had been infiltrated by Communists, a decision was made to clear the veterans from encampments on government property. Two veterans were killed, whereupon President Herbert Hoover ordered the removal of all veterans from the capital. The U.S. Army, under the command of Chief of Staff General Douglas MacArthur, along with Major Dwight D. Eisenhower and Major George Patton, fired tear gas into a crowd that mistakenly thought the display of military might was in their honor. Ignoring Hoover's orders, MacArthur then led his soldiers across the Anacostia River to the marcher's main camp. The ensuing battle resulted in several deaths and hundreds of injuries. Hoover, already seen as incapable of relieving the economic suffering of the Great Depression, appeared cold-heartedly cruel as well. Following the Bonus March, "the war department reviewed the government's treatment of Coxey's army to help the Hoover administration plan for the future."[22]

To Barber, it was a march that never took place—"the Negro March," organized in 1941 to press Franklin Roosevelt to support African American contributions to the war effort—that "firmly established demonstrations in Washington as a distinct and legitimate alternative to voting, lobbying, and petitioning for American citizens."[23] The Roosevelt administration, fearful of the prospect of African Americans marching on the nation's capital and calling attention to inequality while it was negotiating formal entry into World War II, entered into behind-the-scenes negotiations with protest leaders to resolve the issue. When Roosevelt signed Executive Order 8802 to end discriminatory hiring practices in factories and plants engaged in the war effort, planning for the march ceased. African American civil rights leaders had successfully shown that a protest in Washington could no longer be simply dismissed as illegitimate and that even the threat of one might bring about political change.

The 1963 March on Washington

One of the planners of the 1941 Negro March, labor and civil rights activist A. Philip Randolph, is credited with issuing the call for the March on Washington for Jobs and Freedom in August 1963. By the early 1960s, the goal of covering big stories had coincided with a significant and important change in the way many Americans learned about current events—television—and, fortuitously, the civil rights movement took place at about the same time.[24]

In addition, though some Americans had not even heard of the civil rights struggle or calls for integration, television executives seized upon coverage of the civil rights movement as a way to claim relevance. Their interest in the emerging movement was probably not the result of altruism but fear of sanction from the federal agency that issued broadcasting licenses. In a 1961 address to the National Association of Broadcasters, Federal Communications Commission Chairman Newton Minow, appointed by President John F. Kennedy, had called the national commercial television industry a "vast wasteland." Following his address, network executives scrambled to find ways to make commercial television programming meet Minow's call for the public good. By the summer of 1963, the industry had reached multiple milestones centered on the issue of civil rights. Those milestones included an NBC preemption of regular Labor Day programming for a series titled

The American Revolution of '63 and an ABC five-part series that began August 2 titled *Crucial Summer*.[25]

When the call for the march was issued in 1963, the two issues of foremost concern among African Americans were unemployment and racial discrimination. Racial discrimination prevented African Americans from purchasing a home or living in certain communities, attending some schools, shopping in some areas, and voting in some parts of the country. They were also often unfairly targeted by law enforcement. And in 1963, African American unemployment stood at 11 percent—double that of white Americans.

Randolph's discussion with labor leaders and others helping to organize the march produced an agreement on several goals of the event, including passage of the Civil Rights Act, integration of public schools, legislation prohibiting job discrimination, and a call for job training and placement.[26] The Kennedy administration was opposed to the march and cited the potential for violence. But an increasing number of groups signed on, and the march's themes were expanded to include unity and racial harmony.

Bayard Rustin, a march co-organizer, worked diligently to prepare for an estimated participation of a hundred thousand people. But far more—a quarter of million—joined the march on August 27. "Freedom buses" and "Freedom trains" organized by churches and civil rights groups brought marchers into Washington from all over the country. But the march was not only an event for city outsiders. African Americans in D.C. saw the protest as an opportunity to express their desire for equal treatment as well as raise issues unique to the city. Some twenty five thousand Washingtonians would participate in the protest, some holding signs that read "First Class Citizens Can Vote/D.C. Wants Home Rule," and local civil rights leader Rev. Walter Fauntroy—later elected as the city's first delegate to the U.S. House of Representatives in the twentieth century—was the march's coordinator.[27]

March organizers knew that widespread and positive news coverage was essential for the event to have any political influence. The night before the march, the prepared speeches of all the participants were delivered to journalists and the logistics discussed and approved by several city officials. That made covering the march much like following the scripts for game shows, westerns, and other forms of commercial television that Minow had railed about in 1961. The single unsettled issue of the event was the text of the speech to be delivered by John Lewis, the national chairman of the Student Non-Violent Coordinating Committee. Lewis planned to criticize inaction on ending discrimination by saying African Americans would take matters into their own hands and "march through the South . . . the way Sherman did, leaving a scorched earth with our non-violence."[28] Several speakers threatened to pull out of the event if Lewis did not remove language they considered incendiary. As the crowd gathered at the Lincoln Memorial, a committee worked with Lewis "inside a guard station beneath the giant seat of the Lincoln statue" while Rustin reorganized the sequence of events and Lewis rewrote his speech.[29]

Broadcast live by the three major networks and beamed to Europe via the newly launched satellite Telstar, the images of whites and blacks marching, singing, and listening to speeches were seen by millions of television viewers, including many who up to that point had heard only sparingly about the civil rights movement. Coverage began on NBC's *Today* show with a thirty-minute report and continued throughout the day with a two-hour-long recap in the afternoon and a final report at night. CBS carried the speeches live and then aired an hour-long special report.[30] Importantly, broadcasters consistently emphasized the "peaceful" nature of the protest, despite the hot August weather and the contentious issue of integration.

The seven-minute speech delivered by the Rev. Dr. Martin Luther King, Jr. was the last speech of the day. King began with prepared remarks but then spoke extemporaneously. The next day, some major media outlets ignored the speech; *Newsweek* did, for example, and the *Washington Post* highlighted Randolph's speech instead. For other outlets, however, the "I Have a Dream" speech was—as most would now agree—the highlight of the day. Historian Taylor Branch writes that "setting aside his lofty text to let loose and jam" elevated King to the status of "a new founding father."[31]

The aftermath of the march on Washington is in some respects similar to what happened following Coxey's march. In both instances, the concerns of the marchers were ignored by Congress. No new legislation was signed as a direct result of Coxey's Army or the March on Washington. The 1963 event often gets credit for the nation's subsequent willingness to move forward with civil rights legislation. However, empirical studies contradict that conventional wisdom. One study, for instance, found that "the trends toward support for civil rights began in the 1940s, rather than as a response to a few weeks of television coverage." Another found age-related demographic changes explained shifts in attitudes toward civil rights. A third reported that "public opinion did not change because of dramatic events, demonstrations or news . . . instead it had been growing more supportive of civil rights legislation for years."[32] Nonetheless, the 1963 march showed that protests in the national capital could not only be peaceful but inspirational to millions of people.

The Future

In the new millennium, the effort to "control the message" or enlist the help of journalists to spread the message may be made more difficult by changes in the structure of the mass media. Mainstream newspapers, television, and radio now compete for audiences with blogs, websites, and social media. Politically partisan content competes for the attention of the audience in new and different ways. One study noted that, in the current environment, once social protest movements successfully gain the attention of reporters, contemporary journalists tend to ignore the issues that drive the movement and instead critique the tactics.[33]

Despite the challenges inherent in capturing the attention of the press, holding an event in Washington, D.C., is still an important element of many social movements. A relatively recent example occurred on May 1, 2006, when a nationwide "Day without an Immigrant" was coordinated in several cities across the United States. At the time, the U.S. Congress was debating a bill that would have tightened immigration restrictions, built a wall along the Mexican border, and made undocumented workers felons, and the Immigration and Customs Enforcement had recently rounded up more than ten thousand people and their employers.

The purpose of Dia Sin Inmigrantes was to demonstrate the economic importance of documented and undocumented workers with a one-day work boycott. The challenge was to present the argument without activating the fears of those opposed to immigration who claim immigrants take jobs away from American workers. Organizers urged protestors to cultivate the image of peaceful protest in harmony, as used so effectively by the 1963 March on Washington, by wearing white, carrying American flags, making use of terms such as *amnesty* and *immigration reform,* and bringing children to the events. A discourse analysis of nearly three hundred newspaper articles and two television shows about Dia Sin Inmigrantes found that instead of being marginalized, coverage of the events dominated the front page of newspapers, and it was the lead story of local and national television news coverage.[34]

The example of Washingtonians participating in the 1963 March on Washington also serves as a reminder that demonstrations in D.C. are not solely for outsiders. The city is overwhelmingly Democratic, and demonstrations that espouse left-leaning causes—ending the war in Vietnam or Iraq, protesting the unusual election of George W. Bush in 2000, or even a rally featuring Jon Stewart and Stephen Colbert—can count on a large turnout by locals. So too can protests organized around particular ethnic groups that are prominent in Washington. Dia Sin Inmigrantes brought many of the capital's Latinos to the Mall, for example, and despite the title of Spike Lee's film about the Nation of Islam's 1995 Million Man March—*Get on the Bus*—thousands of that protest's participants were Washingtonians who did not need to take a cross-country bus to get to the Mall.

Protests, demonstrations, and rallies are so much a part of the everyday scene in Washington that some have begun to question their usefulness. They point to the lack of interest on the part of many media outlets, which often dedicate no more than a brief segment of pro forma coverage on the nightly news. Organizers are also aware that modern technology makes news coverage from any place on the planet possible; thus, the expense and preparation required for large groups to travel to the nation's capital may be better used in other ways. For a while, the size of D.C. protests were measured against the numbers generated by the 1963 March on Washington and found deficient.[35] The goal of obtaining even larger numbers of protesters has been thwarted by the National Park Service's decision not to provide crowd estimates, which followed disputes over whether a million people truly participated in the Million Man March. Also, since the September 11, 2001 attacks there is now the issue of security: concerns about safety mean that officials have many more checkpoints for participants and additional restrictions on who can go where and how close to the seat of power demonstrations can be.

The empirical evidence of a positive association between a protest in the nation's capital and significant political action is thin. Still, the United States Park Police, the agency charged with issuing permits to organizations and groups that want to stage a protest in the nation's capital, issues hundreds of permits a year. Sometimes the protests draw media attention; sometimes they do not. But either way, protests in Washington, D.C., seem to have become settled practice—and groups of all sizes and persuasions agree that demonstrating the ability to mobilize and organize large numbers of fellow citizens is an ideal way to "speak truth to power."

CHAPTER 6

Political Host to the World

What makes a city international? For some cities, it is having a diverse immigrant population.[1] For others, it is economic interdependence with foreign markets. Both are features of Washington, D.C. But the city is also home to countless embassies, legations, and international governmental organizations (IGOs); dozens of foreign political leaders visit it every year; and it has scores of nonprofits and think tanks that focus on multinational political issues.[2] Taken together, they constitute another side to the city's international character that is shared by few other American urban places: its status as a center of international *politics*.

It was not always so. For many years, the United States was not considered important enough for European nations to establish prominent diplomatic settlements here. The capital was not an attractive place for diplomats to live, either. It was not until the end of the nineteenth century that Washington began to emerge as a necessary, even desirable, place for the emissaries of countries to settle—a development that finally blossomed after World War II, when the United States became the global superpower it is today. The presence of so many international political entities has provided the city with unique cultural and political benefits but has also introduced new complications and challenges for politicians, local residents, and city government.

A Brief History

In the late eighteenth century, establishing an embassy in a foreign country created a powerful imprimatur of legitimacy: it signaled that the country was worthy of recognition and a significant player in world affairs. When the United States was first established in 1788, the new country hardly fit that description. It was largely rural, far from the influential nations of Europe, and its central government was weak and had no standing army. France did send a foreign representative in 1778, before the war for independence was even won (the first country to do so), and England, America's one-time mother country, established the first foreign office in the United States. But England's diplomatic office was technically a legation, meaning that the country did not assign a permanent ambassador to represent it—a sure sign Britain did not consider America as important as the "major" powers of the globe.

Early Washington was not an appealing destination for foreign ministers. Pierre L'Enfant may have hoped that a main boulevard running due west from the Capitol would contain some diplomatic missions, but the boulevard was never built, and the area (today's National Mall) remained undeveloped for decades.[3] Foreign offices were opened elsewhere in the city, and chronicler Frances Trollope observed in 1831 that Washington's diplomatic contingent was what "distinguishes it [the city] greatly from all others." Nonetheless, the capital had a smaller social scene, and fewer theaters and other entertainments, than larger and more established American cities such as Philadelphia, New York, or Boston—let alone the major capitals of Europe. Unhappy was the foreign emissary forced to spend his days in what the secretary to the minister from England in the first decade of the nineteenth century called the "Washington wilderness." Some chose to frequent other places on the continent, coming to Washington only when business demanded it. Spain's minister in the late 1790s "found excuses to spend most of his tour in Philadelphia and New York." The minister from Britain "disliked the new capital instantly" and left for the West Indies, never to return. The representative from the Netherlands in 1801 fled after just over a year of living in the new city.[4]

Disinterest in the United States lasted through most of the nineteenth century. While "occasionally" considered "a possible chess piece in the game of world politics," America was not thought of "as a nation that might become a player of power in its own right." Though Washington became a more attractive place to live following the internal improvements undertaken in the 1870s by Alexander "Boss" Shepherd, it was still far less exciting than other capitals. The city remained bereft of embassies and featured only legations, and even these were often considered to be of little importance compared to those in other countries. Forced in 1880 to close its least valued diplomatic missions in order to save money, for example, Turkey shuttered its Washington office altogether. Furthermore, foreign dignitaries and heads of state rarely came to the United States, a country still too distant and insignificant to pay a visit. Nor did the United States necessarily receive the highest quality diplomats from abroad. In 1871, for instance, the Russian emissary Constantine de Catacazy committed a gross diplomatic faux-pas by trying to sway American policy makers with anonymously published criticisms of the White House and the State Department. Unable to secure Catacazy's immediate dismissal, President Ulysses Grant refused to interact with him or even invite him to dinner with the visiting Russian grand duke; Catacazy was eventually recalled. Thirteen years later, when Otto von Bismarck needed to replace the German minister to America, he selected a minor figure who had previously represented a small principality.[5]

Things finally began to change as the 1880s came to a close. The United States was growing rapidly in size and wealth, had begun building a sizeable navy, and increasingly flexed its diplomatic and military muscles. The world's great powers realized that they needed more prominent representatives in Washington, D.C. England was the first country to upgrade its office in the city from a legation to an embassy, thenceforth sending an ambassador, not an emissary of lesser title, beginning in 1893. Germany, France, and Italy soon followed England's lead, and the United States reciprocated by establishing embassies in their capitals.[6]

As America's power grew in the late nineteenth and early twentieth centuries, additional embassies were established at its seat of government. New international institutions began to appear. The millionaire philanthropist Andrew Carnegie donated funds to build the headquarters of the Organization of American States near the Mall. It was completed in 1910, the same year he created a new institute in Washington, the Carnegie Endowment for International Peace, to encourage the peaceful resolution of international conflicts.[7] The city also became the host of occasional multinational conferences. The Inter-American Conference, the first security conference among nation-states held for reasons other than ending a specific conflict, took place in Washington in 1889, an event one historian called "a milestone in the emergence of a modern, globally connected Washington."[8] Official visits by overseas leaders (usually, though not always, to Washington), originally a sporadic occurrence, began happening at least once a year starting in the 1920s (see figure 6.1). Nonetheless, these visits were few—no more than three annually—and were usually by leaders from the Americas, not Europe or Asia. And Washington was still not especially appealing to diplomats accustomed to larger, more exciting locales in other countries, let alone bustling American metropolises like New York.[9]

World War II brought the nation, and thus the capital, unprecedented international prominence. The United Nations would be headquartered in New York City, but its birthplace was Washington, D.C.—the result of a 1944 meeting in Dumbarton Oaks in Georgetown—and in the aftermath of the war two important IGOs, the World Bank and the International Monetary Fund (IMF), established their headquarters in the capital. Newly independent former colonies made sure to open embassies in the city, often occupying old mansions in the tony nineteenth-century neighborhoods of Dupont Circle and Kalorama. Washington went from having 50 embassies

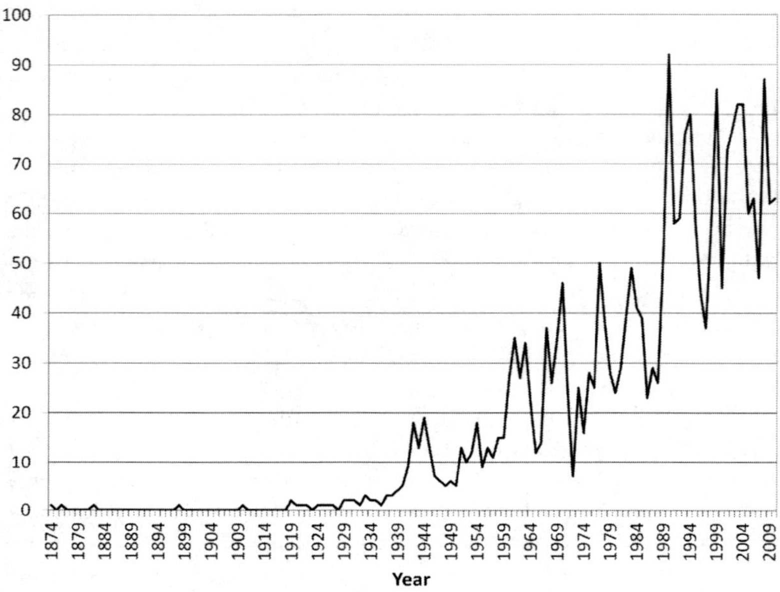

Figure 6.1 Number of Official Visits to the United States by Foreign Leaders
Source: U.S. Department of State, Office of the Historian (http://history.state.gov/departmenthistory/visits)

when World War II began to over 130 in the early 1980s. Easier air travel meant more foreign leaders could visit the country, and more often. Washington, D.C., was usually their primary, if not only, destination. White House state dinners for overseas guests became routine by the 1950s. "World leaders have ceased to be novelties on the capital scene," observed local reporter Hope Ridings Miller in 1969, who also noted how their visits to the city could require the closure of multiple streets. "All this commotion about foreign nabobs sure louses up the traffic!" complained one taxi driver trying to navigate a parade honoring a visiting monarch.[10]

Washington's status as an international political city expanded in a new direction starting in the 1970s. Companies that did business with the U.S. government, particularly information-based ones that saw Washington as a "specialized information city," found the district an ideal place to locate satellite offices. The arrival of these companies involved politics as well as economics, since decisions by the U.S. government on spending and procurement affected the bottom line of a growing number of foreign firms. The effect on the city and region was dramatic. Dulles Airport in northern Virginia transformed from an underused facility in the 1960s to a busy terminal for international travelers within a few decades. Between 1980 and 1991, the number of area companies and organizations with international names more than doubled, from 493 to nearly 1,300.[11] Many of these entities, together with government agencies focused on encouraging trade, opened offices in the mammoth Ronald Reagan Building and International Trade Center, built in 1998 just blocks from the White House.

Today, Washington is a major hub of international politics. There are over 175 diplomatic missions in the city and, as one journalist observed, "Most countries maintain their largest foreign missions in Washington."[12] The District of Columbia's many embassies are joined not only by the World Bank and the IMF but also by another IGO, the Pan-American Health Organization, and a slew of nonprofit entities focused on international affairs, ranging from the German Marshall Fund to the Meridian International Center.

Space, Location, and International Prestige

Locating an embassy or international organization in the capital of the most powerful country on earth makes obvious political sense. But where in the city to locate and in what sort of building are important political decisions too. When these entities have had the freedom to make those decisions, their choices have frequently reflected not only budgetary concerns and issues of convenience but also *prestige*, a central principle of international politics.

In his landmark book on international affairs, *Politics among Nations*, Hans Morgenthau observed that a country's power is shaped in part by how it is perceived by others. Even countries that seem obviously weak—having a small economy or limited military, for example—may, through skillful leadership and diplomacy, convey an impression of power that translates into actual influence. Likewise, nations that seem strong on paper may not be feared or respected by others and have correspondingly less influence abroad. Prestige is thus a cornerstone of ambassadorial politics, since the impressions made by a country's embassy and ambassador may shape, however slightly, that nation's reputation in the eyes of the host country.[13] They are also important for IGOs whose power is greatly dependent on the willingness of other nations to follow their leadership.

International entities in Washington can create positive impressions in two ways. One is through the physical design of their buildings. A large edifice implies power and importance; certain design features can catch the eye and convey strength and authority; and an embassy's novel design may show off its country's home-grown architectural talent. All around the capital one can find examples of the use of architecture by foreign and international political entities to heighten their prestige. There is, for instance, the imposing glass and concrete World Bank headquarters, with its huge indoor foyer, located northwest of the White House. The Canadian embassy's massive marble edifice includes a large reflecting pool and even a small open atrium that amplifies one's voice in a resounding echo (see figure 6.2). The Embassy of Finland on upper Massachusetts Avenue, designed by two Finish architects, is made of eye-catching glass and steel. A good number of embassies are in older urban manors that convey Old World grandeur.

The other way to convey prestige is by location. It is a badge of honor for an embassy to be on the northwest stretch of Massachusetts Avenue known as Embassy Row. The street gained political prominence and its nickname when Great Britain, for centuries the most powerful empire in the world, built a new embassy on the street in the 1930s.[14] Other countries followed suit. Some moved into stately Gilded Age homes along or near the avenue; others, such as Finland—which, its website proudly notes, is "located across from the Vice President's residence"—purposely constructed new facilities on

Figure 6.2 The Canadian Embassy Building

Source: Library of Congress, Prints and Photographs Division, Photograph by Carol M. Highsmith, LC-DIG-highsm-16442

the thoroughfare. Embassy Row is not the only place in the city where one can earn prestige. Other visually or politically prominent buildings include the Canadian Embassy, located near the Capitol and the only embassy on Pennsylvania Avenue; the World Bank, also on Pennsylvania Avenue and close to the White House; the Russian Embassy, on a high point in northwest Washington; and the Swedish Embassy, in the fashionable neighborhood of Georgetown with a view of the Potomac River.[15]

To be sure, the location of an embassy or international agency often serves utilitarian purposes. Being close to other international organizations facilitates communication and work-related travel. A proximity to downtown Washington and U.S. federal agencies is useful for lobbying. Nearby amenities matter too: as one World Bank official put it, "We also get a lot of visitors from overseas so we need to be close to hotels and other services." And there are additional ways for an ambassador to protect or enhance his country's prestige, such as by throwing extravagant receptions for his visiting head of state or ensuring he is seated in the proper location at formal events.[16] Nonetheless, a noteworthy address is extraordinarily useful for any country or agency concerned with its prestige, as most are.

The Opportunities and Hazards of an International Political City

Besides a smattering of red and blue diplomatic license plates, what does it mean for Washington to have so many embassies and international organizations within its borders? On the one hand, being a center of international politics offers some unusual opportunities to the city. One is economic. An estimated 10,000 people work in embassies, and many of them live and spend money in the district. International organizations lease Washington office space and buy food and supplies from local companies. Embassies also draw visitors who rely upon their services, including American citizens needing visas to visit abroad; foreign nationals seeking travel, legal, and other assistance while in the United States; and American companies seeking help to do business in other nations. The latter is especially important: as regional expert Professor Stephen Fuller argues, "It's what makes Washington a magnetic force for businesses who want to work in those 190 countries with embassies."[17] Altogether, international political entities in the district contribute over $400 million a year to the Washington-area economy.[18]

Another unique benefit of having so many embassies and international political organizations in Washington is their contribution to the city's local culture. Public diplomacy—conveying a good impression of a country and educating people about its history and society—is one of their most important missions. Accordingly, many hold cultural events in the capital, including art shows, concerts, food fairs, talks, and receptions. Embassies are not the only institutions that sponsor cultural and social events in Washington, either. The Reagan Trade Building, home of many internationally oriented agencies and firms, also holds an international food and wine festival every year, while the International Club of D.C. sponsors concerts, dances, and even ice skating lessons for "internationally-minded professionals who enjoy international cultural experiences." Tours and open houses of embassies draw tens of thousands of visitors annually. A country's embassy can also transform its host city into a refuge for its emigrants, as some have argued was the case for Ethiopia. Embassies can contribute to the city's local politics and society in other ways, too. A particularly dramatic example—one that had racial implications for Washington—involved the Turkish Embassy in the 1930s and 1940s. The sons of the Turkish ambassador, both huge jazz fans, openly invited black musicians to come "through the front door" of the ambassador's residence to play concerts for invited guests. Complaints from Southern senators of this open violation of the city's unofficial norms of segregation failed to deter the pair. To the contrary, they actively promoted and sponsored jazz concerts elsewhere in Washington. "Jazz," as one of them later put it, "was our weapon for social action."[19]

The social scene of the city's political elite in particular has long been seasoned by the presence of ambassadors and dignitaries from abroad. During the Gilded Age, Washingtonians treasured invitations to balls sponsored by foreign missions. In the days before air travel, ambassadors and foreign dignitaries added an exotic flair to Washington society and were thus in high demand. As late as 1924, one commentator observed that "through the social season" embassy officials, especially the unmarried ones, "lead the lives of the hunted." Even today, ambassadors spend a good amount of their time either hosting or attending dinners or other social events with the capital's hoity-toity. Doing so allows them to show off their country and "spotlight attractions and customs of [the] homeland" in order to heighten its prestige. It also allows them to build and foster connections and maintain good relations with influential Washingtonians. "Being an ambassador these days is a bit like being an airline stewardess," remarked one U.S. ambassador in 1991. "You serve many meals, and you clean up minor messes."[20]

These special advantages to having international organizations in Washington—more dollars spent in the city, many diverse cultural activities, and an influx of foreign visitors and immigrants who contribute to urban diversity and social life—are mitigated by some important and unusual constraints, influences, and even dangers. For one thing, Washingtonians (and national politicians especially) must remain aware of how they behave in the presence of foreign dignitaries, lest the city's reputation, and even the prestige of the United States, is tarnished. Thomas Jefferson learned this lesson as president when, deeply resentful toward the British Empire, he deliberately broke rules of etiquette with England's minister to America, Anthony Merry. Jefferson dressed shabbily during his first meeting with Merry and openly breached proper etiquette with the minister at formal events. Merry's disgust was so great that he sent reports back to London suggesting the United States was not a reliable ally—reports that may have contributed to England's policies toward America that led to the War of 1812. Embassies also regularly inform government officials of their views on domestic policy and, though less common than it once was, try on occasion to influence American politics directly—to the chagrin of policy makers in Washington. Almost a hundred countries hire D.C. lobbyists, and ambassadors and embassy staff do not hesitate to inform government officials of their concerns with policies that might affect their countries, particularly in areas of trade and immigration. Some, such as Israel, Greece, and Ireland, can take advantage of America's ethnic diversity and gin up support for their cause among U.S. citizens with cultural, religious, or familial ties to their country.[21]

Being a center of international politics also forces Washington to handle unusual levels of policing and crowd control, because embassies and international organizations routinely attract protests. While many demonstrations are small and peaceful—sometimes no more than a solitary person holding a sign or handing out pamphlets—some can become large and unruly events. In the 1980s, the South African embassy was a frequent target of large demonstrations against racial apartheid. China's embassy is routinely targeted by people protesting everything from its treatment of Tibetans and Uyghurs to the country's support for authoritarian regimes, such as in Ethiopia. In 2011, locals who gathered in front of the Turkish Embassy to raise awareness of the genocide of Armenians in Turkey in the early 1900s faced counter-protestors who "waved baseball bats and sang and danced in the streets," and local police were deployed to ensure the two groups did not come to blows. When the actor George Clooney was arrested in front of the Sudanese embassy on Embassy Row to protest human rights violations in Sudan, a scrum of reporters and cameramen swarmed the sidewalk along Massachusetts Avenue. The city, which cannot collect taxes on foreign missions or their staff, must bear the brunt of any costs associated with policing these protests.[22]

There is also the potential for a "spillover effect": clashes in other countries spreading to Washington. Usually such conflicts are avoided. For instance, the Tamils and Sinhalese of Sri Lanka have a long-standing and deep-seated enmity for each other, but those who live in D.C. make a conscious effort to keep apart, even when patronizing the same local tea shop. Nonetheless, the city's international status means that, as one Tamil American put it, "Washington is another frontline of these conflicts." In early 1967, the chancery of Yugoslavia was one of several around the country damaged by explosives planted by a Yugoslav émigré who opposed the country's Communist government. The most disturbing instance of a spillover effect occurred in 1976. The Chilean secret police had targeted a former government official in exile in Washington for assassination: Orlando Letelier, an outspoken critic of Augusto Pinochet, the military dictator of Chile. On the morning of September 21, Letelier was driving along Embassy Row when his vehicle was blown up by hidden explosives, killing him and an aide who was also in the car.[23]

Finally, embassies and foreign dignitaries underscore—if not heighten—Washington residents' lack of political influence and autonomy. One source of irritation is diplomats' special legal status and their ability to avoid prosecution for committing crimes.[24] A recent news story revealed that diplomats owed the city over half a million dollars in speeding and parking tickets. And there have been far more serious cases of illegal activity by diplomats and their family members. The son of the minister from Ireland was not prosecuted for a fatal hit-and-run accident in 1959, thanks to diplomatic immunity.[25] In 1997 an embassy official from the Republic of Georgia killed one young teenager and injured four others in D.C. while driving under the influence. The previous year the same official had avoided prosecution for drunk driving after claiming diplomatic immunity.[26] Legal immunity was claimed again in 2005 by a diplomat from the United Arab Emirates. His diplomatic status allowed him to escape arrest in Virginia for soliciting sex over the Internet from a police officer pretending to be an underage girl, and he subsequently fled the United States.[27]

Residents have also found themselves helpless to stop the construction of embassy buildings they find too big, intrusive, or "a blot a city neighborhoods." Though once highly desirable, embassies lost their luster with Washingtonians when they proliferated in number in the 1950s, causing parking and traffic problems. Special zoning rules regarding the construction of new missions were adopted in 1964; it alleviated the problems to some extent, but conflicts between foreign nations and local neighborhoods continued. Embassies often win these conflicts. For instance, when Turkey sought to rebuild and expand its 1928 chancery on Embassy Row, residents of the nearby Kalorama neighborhood who opposed the new design were unable to halt it. The relative weakness of city residents and local government against the federal government is often made crystal clear in these conflicts, too. In 1979, responding to the complaints of residents of both Embassy Row and Sixteenth Street N.W. (another popular street for embassies), the city council passed a law limiting the construction of office buildings used by embassies (known as chanceries). But the State Department pleaded that it "must provide foreign nations with suitable sites in Washington in return for U.S. concessions abroad," and Congress overruled the law—the first time, and one of the only times, Congress has done so since the city was given home rule in 1973.[28]

Conclusion

This chapter necessarily offers but a small sample of the many ways that international politics manifests itself in the city of Washington. Little has been said, for instance, about what most people probably associate with the capital's international status: espionage. There have been several notable instances of spying and covert activities in the D.C. area. During World War II, for instance, a woman named Elizabeth Thorpe Pack seduced key diplomatic figures in the city to obtain classified information from the Axis powers for the British government.[29] Well-placed U.S. government officials including Aldrich Ames, Robert Hanssen, Ronald Pelton, and John Anthony Walker provided Russia with classified information both during and after the Cold War. Some high-ranking officials from other countries, including Yuri Nosenko (a Soviet KGB agent) and Milan Švec (a minister from Czechoslovakia), have chosen Washington as the place to defect from their country and seek asylum in the United States. Though these instances do not necessarily make Washington a true cloak-and-dagger city, espionage and intrigue happens often enough to justify a spy museum in downtown Washington and regular "spy tours" of the area, where tourists are shown famous meeting places and dead drops—places where secret information or material is left—of famous spies of the past.

Foreign intrigue, bold international buildings, a huge number of embassies, cultural influences, protests, and diplomats who lobby, entertain, and occasionally break the law—all are consequences of Washington's standing as a place of international politics. As long as the United States remains a global military, economic, and cultural force, international politics will continue to contribute to the day-to-day affairs of our nation's capital.

CHAPTER 7

Home Rule, Race, and Revenue: The Local Politics of Washington

Washington, D.C., the center of national politics, also has a politics all its own. As in any city, Washington's local politics can be quite complex: community leaders, the local media, businesses, unions, churches, and government officials competing and cooperating to make policy for a diverse citizenry. But three central, enduring, and interconnected elements of Washington explain much of its local politics and make those politics unique. The first is D.C.'s lack of full self-governance and representation in Congress. The second is the influence of African Americans and racial attitudes and identity in determining city affairs. The third is the significant constraints on the city's ability to raise and spend revenue. In this chapter, we examine each of these themes in detail and the ways they distinguish Washington, D.C., from other American cities.

Self-Governance and Congressional Representation

The United States may consider itself a beacon of democracy, but in its own capital city that beacon shines but dimly. The absence in Washington, D.C., of two cornerstones of democratic rule—full self-governance and federal representation—serves as a rallying cry for its inhabitants, a limit on the governing authority of its elected leaders, and an opportunity for the federal government to shape city policy.

Take first Washington's lack of full self-governance. To be sure, like any other American city, the capital has a system of local government. There is a mayor who serves a four-year term and a city council of thirteen members also elected to four-year terms. Together they draft, consider, and enact local laws and an annual budget. To further empower neighborhoods, the city government includes Advisory Neighborhood Commissions (ANCs), elected groups responsible for conveying concerns and suggestions from areas of the city to the mayor and city council. Furthermore, with no separate county government to deal with, D.C.'s local politicians have more autonomy and less bureaucratic entanglements than their compatriots in other urban places.[1]

However, while the arrival of local government to Washington in 1974 was touted as "home rule" for the city, it is home rule of a very limited sort, for Congress retained the explicit power to have the final say on all local laws and the city's budget. In fact, the District of Columbia is the only place in America named in the U.S. Constitution for which Congress is given the right "to exercise exclusive Legislation in all Cases whatsoever."[2] Congress has been far more eager and willing to use that authority than the few national legislatures of other countries with similar powers over their capitals.[3] Table 7.1 lists a sample of the many kinds of

Table 7.1 Examples of Congressional Directives to Washington, D.C., since 1974

Policy Area	Examples (Dates)
Abortion	* No federal funds may be spent on abortion (1979–present) * No city funds may be spent on abortion (1988–1992, 1995–2008, 2010–present)
Gay Rights	* No federal or local funds may be spent to implement city law granting rights to same-sex couples (1992–2002)
Lobbying Congress	* No funds may be spent to lobby Congress for statehood/voting representation in Congress (1998–2005)
Lottery	* Advertising of lottery prohibited on public transportation (1981)
Medical Marijuana	* No funds may be spent to implement city law legalizing medical marijuana (1998–2008)
Needle Exchange Program	* No federal funds may be spent to give clean needles to drug addicts (1998–2008, 2011–present) * No local funds may be spent to give clean needles to drug addicts (1998–2007)
Taxi Cabs	* Use of meters in taxis banned (1974–1986)
Other	* Swimming pool at local high school must be closed after 9:00 p.m. (1974) * Spending on "test borings and soil investigations" capped at a fixed amount (1974–1982) * Portion of street to be renamed Raoul Wallenberg Place (1985) * Referendum must be held to decide whether to establish death penalty (1992) * Capitol Hill firehouse must be kept open (1986–1993)

Sources: Fauntroy, *Home Rule or House Rule?* 76–79; Harris, *Congress and the Governance of the Nation's Capital*, 149–152; Meyers, *Public Opinion*, 29; Schrag, "The Future of the District," 314–315, 355–371; *Congressional Quarterly Almanacs*.

policies—ranging from abortion to the hours of operation of a local swimming pool—over which Congress has used its power to make or change city laws.

The second major issue particular to D.C. politics is apparent to anyone who has seen a Washington, D.C., license plate. Unlike the mottos on other state plates—"The Silver State," "Sunshine State," "The First State"—the slogan on Washington's is far more provocative: "Taxation Without Representation." Introduced in 2000 to replace the innocuous motto "Celebrate and Discover," it succinctly captures a long-standing grievance of the city: its lack of representation within the national government. Washingtonians can vote for president, but their city has no senators and is represented by a single delegate in the House of Representatives with limited voting rights. (There are two non-voting "shadow" senators and one non-voting "shadow" representative elected by the district, whose sole job is to advocate for making D.C. a state.[4]) Whereas the citizens of other American cities elect U.S. senators and House members with full voting power, plus state legislators who "can link up with their compatriots in the state legislature to protect their political interests," Washingtonians cannot. City residents are denied all but a whisper of influence within the institution that controls its fate, and the observation of an early nineteenth-century Washingtonian still holds true: "Every member [of Congress] takes care of the needs of his constituents, but we are the constituents of no one." In fact, Washington is the sole capital of a federal government whose residents do not have full representation in that government.[5]

What explains the curious absence of political autonomy in and representative democracy for the nation's capital? Writing a few years before Washington was founded, James Madison argued that the reasons for Congress's absolute authority over the District of Columbia were self-evident: it kept the national government free from undue influence by the city or state in which the capital resided, and "public improvements" to the city would be "too great a public pledge to be left in the hands of a single State." But Madison did not suggest that congressional authority should translate into a lack of self-governance or representation for D.C. residents. When a state surrendered its land to create the federal district, he predicted, it would "no doubt provide in the compact for the rights and the consent of the citizens inhabiting" that territory, and Madison confidently declared that the future city's residents "will of course" have "a municipal legislature for local purposes, derived from their own suffrages." Nor did Madison advocate a lack of representation in Congress for city residents.[6]

Madison's confidence that there would be democracy for the district was misplaced, however. In 1790, when Virginia and Maryland offered portions of their territory to create the new ten-mile square federal district along the Potomac, both states insisted that their (now former) residents should be represented in Congress and retain their suffrage and other state rights, and

Congress complied.[7] But when the district was officially turned into federal territory eleven years later, no right to participate in federal elections was explicitly granted to its inhabitants—a "historical accident," in the opinion of two legal scholars. Residents of D.C. could not vote for president until 1964 and, except for a brief period in the 1870s, had no representation in Congress until 1970. And though Washingtonians were able to elect at least some members of city government for the first seven decades of the city's existence, in 1874 Congress replaced Washington's locally elected government with a three-member board of commissioners appointed by the president, a board made permanent in 1878 (see table 7.2).[8] For the following nine decades, the only people allowed to govern the district besides the president and congressmen were the city's three appointed commissioners, with informal influence held by members of the (white) business community through its Board of Trade organization.[9]

Table 7.2 Changes to City's Internal Governance and Representation in National Government, 1802–2011[a]

Year of Enactment	City Governance[b]	Representation in National Government
1790	Three commissioners	Citizens retain representation in Congress, right to vote in national elections
1801		Citizens lose representation in Congress, right to vote in national elections
1802	Mayor (appointed by president), city council (elected), upper chamber (appointed by council)	
1804	Upper chamber made elected	
1812	Mayor (chosen by legislature), Board of Alderman (elected), city council (elected)	
1820	Mayor made elected	
1871	Governor (appointed by president), city council (appointed by president), House of delegates (elected)	City given an elected (nonvoting) delegate to U.S. House of Representatives
1874	Temporary board of three commissioners (appointed by president)	House delegate eliminated
1878	Board made permanent	
1961		Right to cast ballots for president in electoral college

(*continued*)

Year of Enactment	City Governance[b]	Representation in National Government
1967	Mayor-commissioner (appointed by president), city council (appointed by president)	
1968	Board of Education (elected)	
1970		City given an elected (non-voting) delegate to U.S. House of Representatives
1973	Mayor (elected), city council (elected), neighborhood commissions (elected)[c]	
1978	Right for citizens to put initiatives, referenda on ballot	
1993		Limited voting rights granted to U.S. House Delegate
1995	Financial control board established with power over certain city operations	Voting rights eliminated for U.S. House delegate
2001	Financial control board disbanded	
2007		Limited voting rights granted to U.S. House delegate
2011		Voting rights eliminated for U.S. House delegate

Sources: Fauntroy, *Home Rule or House Rule?;* Green, *Washington, Vol. 1;* Tydings, "Home Rule for the District of Columbia."

[a] Shaded square indicates change in which the independence and/or representation of the city was curtailed.

[b] From 1801 until 1871, local cities within the district had their own municipal governments; the district was governed by a nine-member, presidentially appointed levy court (Green 1962, 27; Lessoff 1994, 31).

[c] The legislation only required a ballot referendum on whether to create such commissions, which passed the following year.

It took many decades of lobbying by city residents and political leaders before Washingtonians were granted a measure of home rule and national representation in the 1960s and 1970s.[10] But Washington was denied all but limited representation in Congress and less-than-complete independence. The president and Congress reserved the power to appoint all of the city's judges; Congress was allowed to overturn any local law within thirty days; and the House and Senate were permitted to modify or add restrictions to Washington's annual budget.[11] As a result, more than a few observers have suggested that Congress resembles a colonial power with respect to the District of Columbia, ruling it with near absolute authority.[12]

Throughout its history, then, Washington, D.C., has occasionally been allocated some degree of local governance and representation in Congress, but never as much as other cities, and it has often had none at all (see table 7.2). The degree of governance and representation allocated to Washington has often been the result of party politics in Congress. When city residents voted reliably Republican in the early 1870s, for instance, it was a Republican-controlled Congress that gave the city its first ever delegate to the House of Representatives, a position the Republican Senate tried to restore in 1878, but Democrats, who by then controlled the House, would not.[13] Today, with Washingtonians voting almost entirely Democratic, the position of the two parties has reversed. Democrats in the House expanded the D.C. delegate's voting rights when they led the chamber in 1993 and 2009, which Republicans reversed when they took over the House in 1995 and 2011.

At least until recently, a second reason Congress has tinkered with Washingtonians' democratic rights and representation is race. After the Civil War, the city's black population had grown significantly, and antislavery Republicans in Congress stood behind local governance and the right of blacks to vote in Washington. But local whites were rarely supportive of black enfranchisement, and, as Republican lawmakers increasingly lost their passion for pursing racial equality, Congress withdrew that support, temporarily eliminating local government in 1874. Washington's budgetary problems confirmed "widespread assumptions about the incapacity of black men to responsibly use their newfound political power." When the Democratic Party—representing white supremacy and the formerly slave-holding South—won control of the House in November of that year, they were happy to put a permanent end to suffrage in the one-third African American city, and did so in 1878.[14] Even though it cost them their own voting rights, many of Washington's whites supported this move to rob African Americans of political power. Later, during the civil rights movement of the 1950s and 1960s, racially progressive congressmen believed that supporting self-governance for Washington, with a large and growing black population, would be "as much a vote against White supremacy as it was a vote for District autonomy."[15] Meanwhile, Southern segregationist Democrats in Congress strongly resisted all efforts to grant self-government and federal representation to the city, despite the tendency of urban voters and African Americans to vote Democratic.[16] As late as 1972, one Southern Democratic congressman openly worried that home rule meant "Black Muslims" would rule D.C.[17]

One explanation that Congress has given for eliminating or limiting local governance is Washington's fiscal health. In 1871, the House and Senate voted to get rid of the city's elected mayoralty and one of its two legislative chambers following reports of fraud and deficit spending by city officials. All local government was abolished in 1874 in the wake of a widely publicized congressional investigation that revealed corruption and gross overspending under the city leadership of Alexander "Boss" Shepherd (vice chair of the city's Board of Public Works from 1871 to 1873 and territorial governor from 1873 to 1874); Democrats were also emphasizing good government in their quest to win that year's elections.[18] After weathering two scandals in three years, ending city government altogether "seemed to many citizens the solution of an otherwise impossible situation," and the "depth of conviction" that Washington could not manage money responsibly "would linger on for the next eighty years." History repeated itself in 1995 when the city's government, crippled by shrinking revenues and spiraling expenses, was forced by Congress to surrender some of its powers to a presidentially appointed control board. Some congressmen may push for these limits on local democracy because they genuinely "don't want to see the District fall apart on their watch," as one congressional aide put it. But often these problems—which are hardly unique to Washington, D.C.—are used to justify denying self-government to the capital for other, less noble motives.[19]

Finally, and perhaps most fundamentally, Congress does not grant Washington greater self-governance or representation because lawmakers like having the power to influence city policy. It is an old temptation that goes back at least as far as the 1830s, when Jacksonian Democrats in Congress took the liberty of "intervening in local affairs, mainly disputes over the financing of capital projects." Some lawmakers—and presidents, for that matter—have interfered out of genuine concern with the welfare of the city, and, as Madison observed, the national government has an inherent interest in ensuring it can conduct its business in the capital.[20] In the first several decades of the nineteenth century, annual presidential addresses often contained a plea to Congress to fund internal improvements in Washington, and President Ulysses Grant (1868–1876) was particularly tenacious in advocating for local development projects. But legislators act out of more selfish motives too. Before the establishment of local government in 1974, members of Congress's D.C. committees appointed friends to positions in city government, and some even used inside information about future projects to enrich themselves.[21] Congressmen have also been known to act purely out of spite. One allegedly pledged to vote against the city's budget when his car was towed away, while another proposed revoking D.C. autonomy completely after his complaints about potholes on a bridge had gone unanswered. Then there are lawmakers from neighboring states who try to shape city laws to help their own constituents at Washington's expense. The city's first convention center, constructed in the early 1980s, was built smaller than intended, in part because legislators representing Maryland feared a bigger center would compete with a similar facility in their state. The shrunken D.C. convention center was soon dwarfed by centers built in other cities and was demolished two decades later.[22]

Presidents and members of Congress also like the opportunity to impose their own, their party's, or their constituents' ideological preferences on the city. History offers plentiful examples. Deeply opposed to the institution of slavery, Republicans eliminated it in the district in 1862 over the opposition of many whites in Washington. Anti-liquor activists in Congress banned all liquor sales in the district in 1917, two years before national prohibition. President Harry Truman used his authority to expand the rights of blacks in the city, part of his push for the Democratic Party to embrace civil rights.[23] President Kennedy's secretary of the interior, Stewart Udall, gave an ultimatum to the Washington Redskins, the last all-white professional football team: draft black players or face eviction from your stadium. Richard Nixon, a conservative law-and-order president, pushed for a strong anti-crime bill limiting the rights of criminal suspects in the city. Today Congress routinely adds language to the city's budget bill banning funds for abortion and needle exchange programs (see table 4.1), and a recent bill to give the city's delegate in the House full voting rights was derailed when Republicans amended it to require that the district repeal its handgun laws.[24]

Many of these efforts are charitably described as "tests" or "experiments" of national policy on a smaller scale. More frequently, they constitute cynical moves to win support from like-minded voters in home districts or from interest groups who see the district "as a model for the states."[25] But sometimes they reflect the interests of certain groups *within* the city who, unable to win at the local level, appeal to Congress instead. When in 1981 Congress overturned city law legalizing sodomy, Washington clergy were among those who had lobbied for Congress to do so. Congressional Republicans reinstated a voucher program in 2011 that allowed district students to go to private schools, pointing to polls showing the program was popular in the city even if it was opposed by local elected leaders.[26]

The disproportionate influence of Congress over the district has had three consequences for the city's local politics. First, it forces D.C. officials to cultivate good relations with members of Congress and consider their policy preferences before enacting new laws, lest Congress retaliate by overturning those laws or otherwise making their lives difficult. As former lobbyist for the city Julius W. Hobson, Jr. put it, "We live in a goldfish bowl, surrounded by members of Congress." In 1989, for example, the city opted not to enact a strict gun control ordinance when a congressman threatened to invalidate it. More recently, Mayor Vincent Grey assiduously courted key congressmen who, despite being from the opposite party, would later prove to be supporters of greater autonomy for the district.[27]

Second, greater independence for the district is a powerful and perennial political cause in the city. Local groups, such as DC Vote, Committee for the Capital City, and Stand Up for Democracy, organize letter-writing campaigns, coordinate public demonstrations, distribute lawn signs, and sponsor other activities on behalf of more autonomy and representation for the city (see figure 7.1). Unsurprisingly, it is "a mantra for every elected official" to make D.C. its own state because it can "always serve as a rallying cry by satisfying emotional needs as well as promising long-denied fundamental political rights."[28] Local elected officials have occasionally gone further than mere rhetoric: Mayor Vincent Gray, for example, was among those arrested in April 2011 during a protest on Capitol Hill after Congress agreed to a federal budget that put limits on the city's ability to pay for abortions.[29]

The third effect of the city's relationship with Congress—a result of the second and somewhat at odds with the first—is that local leaders can gain political points for being perceived as standing up against Capitol Hill.[30] This is especially true when Congress is haughty, arrogant, or paternalistic toward the district—be it the 1980s congressman who boldly declared that "we are the local government here"; the congressional lawmaker who announced that residents who want more independence "can move to Maryland or Virginia"; or the representative who warned that "we gave it [home rule], and we can take it away." One district politician who benefited as much as anyone from

Figure 7.1 Standing up for Greater Rights

this perception of congressional arrogance was Marion Barry, Washington's longtime African American "mayor for life," who in many ways personified black politics in the city from the mid-1970s to the mid-1990s. Barry is a remarkable character and, in many ways, an icon of city politics: charismatic, flamboyant, and often divisive, as Mayor Barry was able to win reelection even in the face of controversies surrounding his governance of the city and his own personal peccadillos—in part by rallying the support of Washingtonians disgruntled by their treatment from Congress. Asked why he voted for Barry as mayor in 1994 despite Barry's conviction and imprisonment for drug possession—an election which Barry won—one resident explained that by doing so he would "at least send somebody up there who they [congressmen] won't push around."[31]

Mayor Barry was, in fact, generally cautious about challenging Congress, aware of its power over the district. His criticisms of Capitol Hill were usually guarded, and he frequently sought common ground with congressional lawmakers.[32] But Barry did sometimes use racial language to rally for personal support. ("I'm not going to be lynched," he proclaimed in the face of a federal probe into accusations of drug use.) He also occasionally voiced complaints that tapped into a deeper fear—widely shared by black voters at the time—that the (predominantly white) Congress might try to "retake" the city. This brings up the second important element of Washington's politics (which we also touch upon in more detail in the next chapter): the importance of race.[33]

Race and Racial Politics

Washington has always had a significant, at times influential, African American population. Benjamin Banneker, an assistant to Pierre L'Enfant's chief surveyor Andrew Ellicott, was a free black man who lived in nearby Maryland, and in the decades leading up to the Civil War the city's freedmen population was one of the largest in the country, though it had little political influence and often suffered considerable discrimination. With the war and the end of slavery, Washington's black population grew from 18 to 33 percent of the city's residents, and congressional Republicans gave them the right to vote in 1866. African Americans began winning seats in the local legislature, participated actively in Republican "ward clubs" and political parades, and helped pass citywide antidiscrimination laws.

Their political influence disappeared, however, when the district lost its ability to elect its own government or representatives to Congress in 1874, leaving power in the hands of those white elites who had the resources and connections (including members of the Board of Trade) to influence Congress informally. Washington's blacks tried to exercise influence through nongovernmental organizations such as the National Association for the Advancement of Colored People (NAACP), which opened its D.C. branch in 1913; the D.C. Federation of Civic Associations, founded in 1921; and the Washington Urban League, established in 1938. But racism and discrimination were major impediments to black empowerment. Even worse, the chairmen of the congressional committees who oversaw the district were usually racist white Southern Democrats. They included Sen. Theodore Bilbo (MS), who once complained about a "black cloud" of African American government workers, and Rep. Ross Collins (MS), who explained that he could not support funds for a D.C. school for impoverished black girls because "my constituents wouldn't stand for spending money on niggers."[34]

The political equation finally began to change for African Americans in Washington (and in other cities) with a major influx of Southern blacks to the North after World War II and the wave of civil rights activism that followed. When Washington became the first non-Southern city to become majority black in the 1950s, "the issue of District voting rights became fused with the cause of civil rights." The number of African Americans in positions of power in D.C. slowly grew. In 1961, President John F. Kennedy named the first black commissioner of Washington, and six years later President Lyndon Johnson appointed an African American, Walter Washington, to be the city's first mayor in nearly a century. Meanwhile, people from around the country were coming to D.C. to fight for home rule, including a young Marion Barry, whose activism became the basis for his rise to power. Another local leader and future D.C. delegate to the U.S. House, Walter Fauntroy, took advantage of newly registered blacks in the district of Rep. John McMillan (D-SC)—an opponent of home rule with authority over district affairs—and led a campaign that ousted McMillan in a 1972 primary election. When Congress first allowed the city to elect its own school board in 1968, seven of the board's eleven seats were won by African Americans, and after Washington, D.C., was finally granted home rule in 1973, black candidates often won, and have continued to win, local elected office.[35]

If it is still true, as political scientist Michael Dawson wrote in the 1990s, that "African-American politics, including political behavior, is *different,*" how does Washington's large, black population uniquely shape the city's politics? For one thing, because blacks vote overwhelmingly for Democrats, liberal candidates have a huge advantage in local elections. So too do black candidates, given that African Americans have traditionally been more inclined to vote for blacks running for office—a consequence of their higher "group consciousness" that mimics the self-perception and voting behavior of ethnic urban voters of the past.[36] This same strong group identity may also make black voters highly loyal to, and thus more hesitant to criticize, African

American officials. Leaders like Marion Barry have in turn been accused of exploiting this race consciousness in order to "rally black voters to their side in reelection campaigns."[37]

In the city's early decades after home rule, the twin legacies of racial discrimination and Congress's dominance over city affairs reinforced some of these race-oriented political tendencies. Without a tradition of local governance, some argued that D.C. residents—particularly African Americans who had recently migrated from a repressive South—suffered from an underdeveloped democratic culture and a lack of community spirit. This in turn may have contributed to a blind faith in city leaders, even those embroiled in accusations of incompetence or corruption such as Mayor Barry. In the 1980s, many in the city's black community came to believe in "the Plan"—a rumored conspiratorial effort by whites to retake control of Washington—which reinforced loyalty to African American office holders and a perception that persecuted leaders like Barry were "yet another link in a long chain of black leaders systematically discredited by racist law enforcement officials." The economic and social problems that were especially prevalent among the city's poor blacks, another unfortunate legacy of racial discrimination, may have helped Barry win elections, especially in his later, more troubled years in office. After being caught on videotape smoking crack cocaine in 1990, some speculated that Barry retained the allegiance of black voters who "had experienced falls similar to Barry's."[38]

However, the voting patterns of city blacks and electoral successes of Washington's African American politicians cannot be explained by skin color alone. For one thing, city residents, regardless of their ethnicity, have voted most loyally for elected leaders who provide them with concrete goods. Barry followed a long-standing tradition of "ethnic" mayors of the past, in cities like New York and Boston, who distributed benefits and services to voters from the same racial or ethnic background and won their support in the process. He provided city jobs to African Americans, sponsored a summer work program for youth (both black and nonblack), brought libraries and trash collection to previously neglected African American neighborhoods, and secured city contracts for minority-run companies.[39] By contrast, a more recent African American D.C. mayor, Adrian Fenty, lost his reelection in 2010, in part by hiring nonblack candidates to leadership posts in city government and allowing his controversial school administrator, Michelle Rhee, to fire a large number of (mostly African American) teachers.[40]

Political skills also matter a great deal. "City politics is a sweaty contact sport," as two D.C. reporters once aptly observed, and candidates like Barry and Fenty won citywide election by being open, accessible, and friendly with everyone. One former city official put it this way: "Marion Barry was the ultimate politician . . . and that man knew *everybody.*" Barry was particularly good at building connections with key Washingtonians, especially church leaders in the African American community who could sway large blocs of voters. He famously called himself "a situationist," able to modify his rhetoric and positions depending on the audience and the political context, and he was masterful at converting former enemies to his side. Victorious candidates do not simply expect their ethnic background to deliver them to office; they must also be able to fund and organize an effective election campaign, as Barry, Fenty, and other successful local politicians in the past did remarkably well.[41]

Nor is race either a consistent or predominant dividing line among city voters. Figure 7.2 shows the percentage of the vote received by the winning mayoral candidate in selected primary elections from two city wards: Ward 3, which is predominantly white, and Ward 8, which is predominantly black. In 2006 Adrian Fenty did well in both wards, and in his first run for mayor in 1978 Marion Barry actually received more of the vote from Ward 3 than from Ward 8. Income is another, often overlapping political cleavage in the city. Ward 3 is wealthy as well as white, whereas Ward 8 is the city's poorest ward.[42] In some D.C. mayoral elections, the winning black candidate has lost support from wealthier voters, irrespective of race.[43]

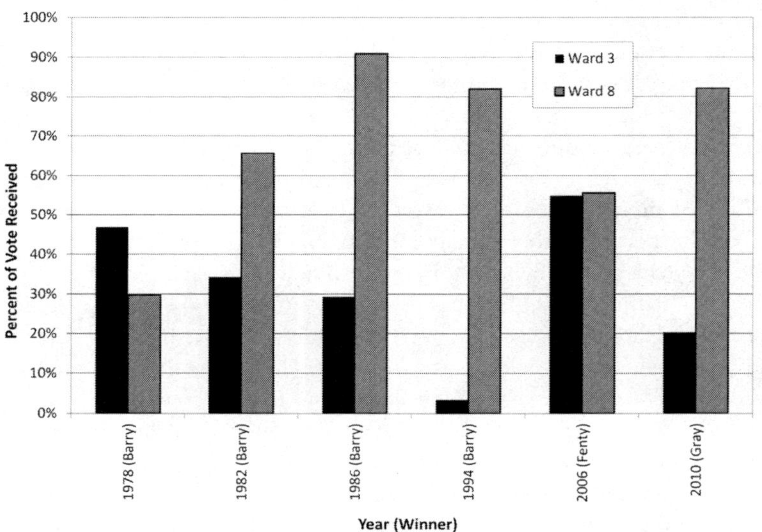

Figure 7.2 Percent of Vote Received by Winning Mayoral Candidate in Wards 3 and 8, Selected Primaries

Finally, racial politics in Washington and other cities have changed and continue to change. Barry was part of a generation of black "civil rights" mayors elected in the 1960s and 1970s that focused on steering jobs and patronage to African American voters and targeting police brutality. But his successors, Anthony Williams (1999–2007) and Fenty (2007–2011), were from a new cohort of "technocratic" black mayors who had also emerged in cities like Newark, Cleveland, and Atlanta "promising to deemphasize race, promote efficient government, and offer strategies to lure investors to strengthen downtown businesses and create jobs."[44]

Budget Politics

Cities face two basic political challenges related to money: where to get it and how best to spend it. Washington, D.C., is no different than any other city when it comes to the second problem. Every year, different groups, interests, and government agencies compete for limited dollars to pay for schools, roads, police, garbage collection, and a host of other city functions and entities. Most of these services are both essential and expensive, and many are dedicated to the nation's neediest—its poor, sick, unemployed, and homeless—who are usually concentrated in urban places.[45] If a city can raise more money, of course, this challenge can be alleviated, if not avoided. But Washington not only faces the same limitations in raising revenue as do other American cities but has additional constraints that further complicate its budget politics and reinforce Congress's power to dictate city policy.

A strong local economy brings with it thriving companies and educated, tax-paying citizens—and thus more revenue—so not surprisingly, "cities constantly seek to upgrade their economic standing." But cities compete against each other for wealth, and all depend on the state of the national and international economy, over which they have virtually no influence, for success. Furthermore, they usually cannot tax urban property owned by the state or federal government, and by one estimate between and four and five times as much tax-exempt property exists in cities as it does outside them. Meanwhile, those who use urban services often commute from the suburbs and thus have no taxable property within city limits. As a result, cities are forced to rely on funding from state coffers and revenue sources such as user fees and taxes on local sales that are "extremely sensitive to variations in the economy," as two urban scholars note. For this reason, writes political scientist Paul Peterson, "city politics is limited politics."[46]

Peterson's description fits Washington, D.C., to a tee. Though the federal government is a large and well-funded employer in the capital, it also owns a lot of city territory that cannot be taxed.[47] The greater D.C. metropolitan area is home to over five million people—one of the most populous in the country—but over 80 percent of them live outside the city's borders and pay income taxes in other states. Even worse, Congress prohibits Washington from imposing a commuter tax (a prohibition staunchly supported by representatives from Maryland and Virginia), while other American cities often collect taxes from workers who live in different states. Washington's local leaders have thus faced the same, if not a greater, challenge to find sources of revenue as officials in other cities have, to the point that they may at times sacrifice the interests of D.C.'s poorer residents in order to entice developers to invest in the district.[48]

The nation's capital has an additional burden related to budgetary politics: its fixed boundaries. In 1890, the district was geographically larger (68 square miles) than such cities as New York (44 square miles), Baltimore (30 square miles), Los Angeles (29 square miles), and Detroit (22 square miles). However, as other American cities grew to absorb outer towns and unincorporated areas, D.C.'s borders remained set by the Constitution and federal law. After four decades, Washington was still 68 square miles, whereas Baltimore's size had more than doubled (to 79 square miles), Detroit and New York had expanded six-fold (to 138 and 299 square miles, respectively), and Los Angeles had grown fifteen-fold (to 440 square miles). These and other cities eventually faced resistance to additional growth from suburban communities, but by then Washington was already far behind in geographic size.[49] Ranked 110th in square mileage among all American cities in the year 2000 with at least 100,000 in population, Washington is forced to operate with relatively little space to house new tax-paying residents and companies.

Thus, D.C. could not even dream of redrawing its borders to recapture the many Washingtonians who moved away after World War II. Americans moved away from plenty of urban places, thanks to cheap suburban housing, ease of driving, and "white flight." Though the exodus was steeper and financially more onerous to government coffers in other cities, Washington's local leaders could still nod their heads in agreement when Philadelphia mayor Richardson Dilworth famously warned in the 1950s that "the suburbs are becoming a white noose strangling the cities." With desegregation, "white flight" soon turned into "wealth flight," as blacks fled to the suburbs as well. In 1970, more people lived in suburbs than in cities for the first time in U.S. history, and D.C.'s population had fallen more than 5 percent from its peak of 800,000 two decades before. By the year 2000, only 570,000 people lived in D.C.[50] Other cities facing the same decline in population could, at least in theory, turn to their states to meet their funding needs, but D.C. had no state government to turn to.

This inability to keep, let alone expand, a large population of tax-paying residents exacerbates political conflicts among different interest groups competing for a piece of the same shrinking budgetary pie. It also pits Washington against its neighboring states that benefit from growing D.C. suburbs, and it ensures that collecting money from commuters remains high on the city's political agenda. And it exacerbates the often-difficult relationship between the city and Congress: for while Washington is the only American city to have consistently received supplemental funding from Congress, that aid comes with significant strings attached.

The U.S. Congress has been willing to intervene in district affairs almost from the very beginning; but, when it began providing significant funding for the city in the 1870s, "taxpayers of the United States were to consider the District budget and administration as no mere local question." Congress usually exercises its influence over city policy through D.C.'s annual spending bills—bills that must be passed for the city to have any budget at all, let alone supplemental dollars from the federal government. That control, and the city's dependence on Congress's largess, was codified in Washington's 1973 Home Rule Charter, which explicitly barred the capital from collecting revenue from certain sources, including commuters and government property.[51] As a result, the city has to impose higher taxes than the typical state, though in comparison to the average city its tax rates are fairly low.[52]

Further hampering Washington, D.C.'s ability to attract new investments is Congress's more-than-occasional reluctance to actively serve as city "booster," especially in comparison to how other states treat their cities. Commitment by the national government to fund Washington's needs has waxed and waned over the course of history.[53] To be sure, federal investment in museums, the Metro, and other attractions and infrastructure projects has helped the capital tremendously. And Congress has been willing from time to time to rescue Washington from bankruptcy, financial mismanagement, and desperately needed improvements to its infrastructure. The 1874 and 1995 budgetary crises that hit the city were resolved when Congress provided Washington with monetary support. But keep in mind that these crises were also followed by Congress taking away some or all of the city's political independence. In both cases, it is telling that many Washington residents were willing to give up their political power in part because they hoped that Congress would provide economic stability to D.C.[54] In this way, Washington's fiscal problems are inexorably tied to the power of Congress and the city's ability to rule itself.

Conclusion

Home rule, race, and budgetary conflicts help explain a great deal of Washington's local politics. Of course, there are other important features of the local political scene. Voters care about jobs and crime, for example, and elected officials who fail to do enough about either may face punishment at the polls.[55] Another, more unfortunate feature that deserves mention is the occasional case of corruption and government mismanagement. Though hardly unique to Washington, political corruption is especially problematic for the city, because many see the capital as representative of the nation and because the failings of African American city leaders may play into existing racial biases.

Corruption and criminal misbehavior became a particularly troubling feature of local politics under Mayor Marion Barry. Stories of gross incompetence and poor governance tarnished the city's reputation: city ambulances, which became lost while answering emergency calls, failing to get patients to hospitals before they died; an understaffed and overworked police force battling drug-related urban violence; and chronic failures to distribute proper housing and educational services to the city's neediest residents.[56] However, it would be wrong to assume that Barry's departure meant an end to city corruption. In 2012, city councilman Harry Thomas, Jr. was convicted of embezzling funds from the city; council chair Kwame Brown resigned amid charges of fraud; scandal-plagued councilman Michael Brown lost reelection; and Mayor Vincent Gray was plagued by ongoing allegations of illegal campaign activity. It is a sad reminder that the potential for corruption is always present, irrespective of who is in office.

Whither the future of district politics? Perhaps the development with the biggest potential to dramatically alter the city's political life is the decline in the percentage of blacks in the city. The 2010 census put the percentage at just a hair above 50 percent, and though D.C.'s white population is growing, there has also been a gradual rise in the Asian and Latino population. Many Washington blacks fear that this may end their political influence, or at least force members of the African American community to join forces with like-minded members of other ethnic groups if they want to continue to shape city affairs. One early sign that changing demographics—both in race and in age—may change the city's electoral politics appeared in a special election for city council in April 2013. Though the incumbent black candidate won, a little-known white candidate and self-described reformer named Elissa Silverman did surprisingly well, losing by only five points and garnering significant support not only among Caucasian voters but younger black D.C. voters too.[57]

Though this trend suggests that racial and ethnic politics could change in Washington, D.C., any shift in the city's budget politics, home rule, or congressional representation seems far more remote. The district has no foreseeable new sources of revenue to draw from, and Congress has every incentive to maintain its influence over city affairs. Attempts to pass a law or amend the Constitution to give the district voting representation in Congress or even convert it into a full-fledged state have also failed. In April 2013 city voters did overwhelmingly approve a referendum allowing Washington to pass its own annual budget without Congress's approval, but its constitutionality was in doubt. One study showed that people endorse D.C. independence when they perceive the city to be economically stable and well-governed; but even so, national support for city statehood has never been high. It would most likely take a nationwide, grassroots campaign to convince American citizens—and, by extension, their representatives in Congress—to fundamentally change the status quo. Until then, as one scholar put it, the city will likely remain "mere plankton in the political sea."[58]

PART III

Washington as Living City

Introduction to Part III

Cities are human constructs. In this respect, it makes some sense to think of a city—where people live, work, and play—as a living place. And certainly Washington, D.C., is no exception.

The last three chapters of this book examine some features of the capital as a living city. These chapters underscore the fact that Washington is much more than a tourist attraction and the home of the national government. As many Washingtonians like to say, their city is both "Washington"—the political capital of the country—and "D.C.," home to tens of thousands of people who have little, if anything, to do with the federal government. It is in these chapters, too, that one sees how the city of Washington differs in important ways from the Maryland and Virginia suburbs that make up the greater Washington metropolitan area.

We start in chapter 8 by looking at immigration to, and ethnicity within, the D.C. region. For most of its history, Washington, D.C., has been a major hub of African American life, with profound consequences for local (and sometimes national) culture and politics. The city and suburbs have also been an important magnet for foreign immigrants, including Europeans, Asians, and Latinos, which has led to a considerable degree of ethnic diversity in the metropolitan area.

In chapter 9 our subject is economics. The conventional wisdom is that Washington is a "company town," heavily reliant on government spending for its economic survival. The truth, however, is more complex. While the city and its suburbs do benefit from federal largess to a greater extent than other parts of the country, the regional economy has greater diversity than many realize, helping to make the area one of the wealthiest in the country. At the same time, the considerable influence of government spending on the local economy—as well as competition between the city and suburbs for business and a steady growth in economic inequality—also creates some significant economic challenges for the city and region.

Our final chapter is about neighborhoods. All cities have neighborhoods, and Washington is no exception. But an examination of the city's neighborhoods—as well as some of its suburban communities—not only uncovers the "living" dimension of Washington, but also shows how some of the city's unique features (including its large African American population, lack of autonomy, and fixed borders) have shaped its neighborhoods and outlying communities in unusual and intriguing ways.

CHAPTER 8

Chocolate City, Vanilla Swirl, or Something Else? Race and Ethnicity in City and Region

"Blacks' Majority Status Slips Away." So proclaimed the *Washington Post*'s above-the-fold headline on March 25, 2011, trumpeting an event that caused a stir among columnists, bloggers, and long-time city residents. The percentage of Washingtonians who were African American had dropped to just half, according to the U.S. Census. It was the lowest level in over five decades and a dramatic change in the city's demographic identity. In the 1970s the capital had earned the nickname "Chocolate City," but as the head of the Greater Washington Urban League put it, "Now they are calling it Vanilla Swirl."[1]

In fact, Washington was known as an epicenter of African American life and political activism long before blacks made up a majority of its residents. The antebellum city was home to both many free blacks and a vibrant slave trade, and some of the former participated in efforts to end the latter. Slaves newly freed during the Civil War—and, later, blacks fleeing Southern segregation—took refuge in Washington, the nearest non-Southern urban center, and the capital's African American population continued to grow, eventually becoming the largest of any major American city. Black Washingtonians pressed doggedly to protect and expand their civil rights and made the city a vibrant hub of African American culture, business, and learning for over a century.

The same census data revealed a second important demographic story. Blacks were not simply being replaced by whites, though the city's Caucasian population had grown by the largest percentage increase in a decade (31 percent) since the 1930s. A variety of other ethnic groups, including Asians and Latinos, were also becoming increasingly sizable in D.C. In other words, rather than resembling a "vanilla swirl," the city was acquiring a more polyethnic, "Neapolitan"-flavored population (see figure 8.1).[2] Yet neither was this an entirely new development. Though never an immigrant city of the same scale as New York or Boston, Washington—home to the national government, a source of secure employment and the basis for economic stability and growth—has been attractive to foreign immigrants since the early 1800s, giving the city a greater degree of cultural diversity than most people realize.

In this chapter we discuss these two important subpopulations of Washington, African Americans and foreign-born immigrants, and how they have changed over the course of the city's history. Both have contributed to the social fabric of Washington in many ways, including the creation of important community institutions, and have occasionally taken advantage of being in the nation's capital to secure rights and liberties for themselves or their brethren outside the city. In addition (as noted in the previous chapter), racial attitudes toward African Americans in particular have been a key factor in determining how Congress governs the city and the extent to which it allows the city to govern itself. Although *race* has little biological meaning, as a social construct—a way that people categorize themselves and others—it has had tremendous power in dictating who may or may not enjoy the full benefits of American society.[3] In Washington, prejudice against blacks helped convince the federal government to end local democracy in the 1870s, just as previous sympathy for blacks had led the abolition of slavery in D.C. and the

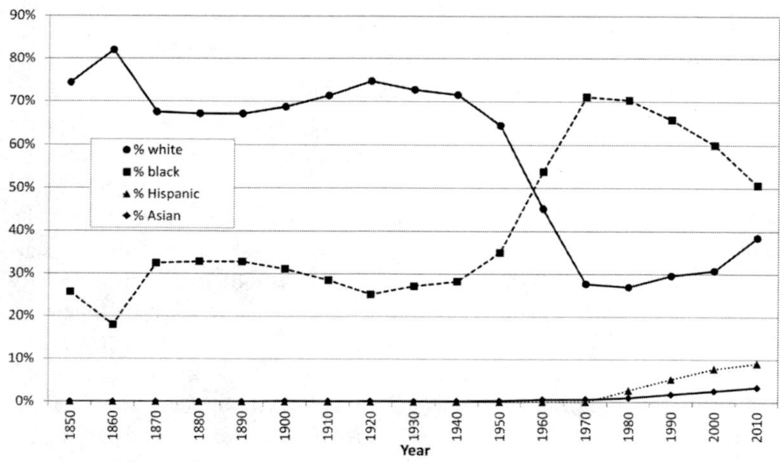

Figure 8.1 Population of Washington D.C., Select Racial/Ethnic Categories
Source: Gibson and Jung 2005 (1850–1990); U.S. Census, Population Division (2000–2010)

expansion of their civil rights, and more tolerant racial attitudes in Congress in the 1970s led to the restoration of home rule for Washington.

Black Antebellum Washington

When first established, the location of the District of Columbia—below the Mason-Dixon Line and surrounded by two slave states, including one (Virginia) with the largest slave population in the country—ensured that it would have a substantial African American population.[4] In 1800, the district was home to over 2,000 black slaves, a quarter of its entire population. Slaves helped build the Capitol and other early government buildings, and several presidents kept slaves at the executive mansion. D.C. also boasted a rapidly growing number of free blacks. Only 400 freedmen lived there in 1800, but the population grew more than tenfold over the next two decades, and by 1860 there were over 11,000 in the district—nearly four times the number of D.C. slaves—making it one of the largest free black urban populations in the country. Some were runaway slaves; others had purchased their freedom; and still others were freedmen who moved to the district by choice or necessity.[5]

Slaves who lived in the city of Washington had some advantages over their counterparts in other states. Working conditions and living arrangements were often more flexible and lenient, and there were several ways slaves could legally obtain their freedom, such as earning enough money by working for others to buy themselves out of slavery. But slavery was still slavery. Nor was life easy for freedmen, who were subject to a series of "black codes" passed by the city beginning in 1808. These included a 10:00 p.m. curfew, limits on where blacks could congregate, and a $500 bond for any new black resident guaranteed by at least two whites. Freedmen also had to carry proof of city residency at all times, and even that was sometimes not enough to prevent forcible kidnapping by slave traders. In 1848, for instance, a black waiter named Henry Wilson was kidnapped and sold into slavery as a spiteful retaliation against his friend Joshua Giddings, a vocal abolitionist congressman from Ohio. (Five years later, Solomon Northup would recount his own kidnapping in Washington and years in servitude in the memoir *Twelve Years a Slave,* an influential book and, a century and a half later, the basis for an Academy Award-winning movie.) A number of city schools, churches, and neighborhoods were segregated by race as well. Though it was possible for freedmen to obtain employment and even to own property, slaves and European immigrants were often given priority in hiring. By 1860, the "[free] black community teetered on the edge of economic and social survival."[6]

During the 1820s and 1830s, the southerly location and lenient business climate of Washington, Georgetown, and Alexandria made D.C. the location of choice for the region's slave traders, and human auction sites sprouted up throughout the district. Slaves were kept and auctioned within blocks of the Capitol building, painting for many an ironic and unsettling image. Just west of the Capitol sat the St. Charles Hotel, built in 1820 with special holding pens where slave-holding guests could store their human property overnight.[7]

Abolitionists had targeted slavery in Washington from its beginning, seeing it as akin to a national endorsement of human bondage. But as the capital became what one abolitionist called "the great Man-Market of the nation," its biracial "subversive community" of local antislavery activists became more vocal and daring. It included Charles Torrey, a former New England minister who helped set up a smuggling operation for escaped slaves in the 1840s, assisting hundreds fleeing north through Washington to freedom before he died of illness at age thirty-two while in jail for his efforts.[8] Another abolitionist, William Chaplin, took over from Torrey and in 1848 helped orchestrate one of the largest attempted slave escapes in the country. Over seventy slaves were covertly put on board the *Pearl*, a ship docked at the Potomac, and sent downriver. Unfortunately for them, the ship was soon captured, and angry pro-slave Washingtonians rioted for three days in protest.[9]

Such violence by locals was not an uncommon danger for district abolitionists and city freedmen. Some acted violently out of loyalty to slavery; others, from pure racial hatred; others (Irish immigrants in particular) because they saw an embrace of anti-black fervor as a means of demonstrating they were of the "white" race.[10] In 1831, the Turner slave rebellion in Virginia led whites in the city to attack African Americans in retaliation. Four years later, hundreds of immigrant workers attacked black homes, schools, churches, and businesses—the so-called Snow Riot—after hearing that a drunken slave had threatened the wife of prominent architect William Thornton.[11]

As Congress became radicalized by the nation's growing divisions over slavery, it also interjected itself into the local debate over the "peculiar institution." One the one side, pro-slavery congressmen worried that ending the slave trade in the nation's capital would, like falling dominos, cause its abolishment in nearby states too. They made sure that slave-state lawmakers dominated committees that supervised D.C., and they even helped pass a new, tougher fugitive slave law to undercut Washington as a stop on the Underground Railroad.[12] On the other side were legislators like Congressman Giddings, who not only delivered a speech in the House of Representatives calling for slaves to rise up against their masters but "challenged propriety in Washington" by admitting African Americans to his capital home as equals and listening to their pleas for help. To the Compromise of 1850, a law designed to resolve the slavery issue, anti-slave lawmakers managed to add a provision banning the sale of slaves in the district from one state to another. Four years earlier, white residents of D.C. who lived on the west side of the Potomac River (including the city of Alexandria, an important slave port) had successfully persuaded Congress to retrocede their region to the state of Virginia—in large part from fear of just such a congressional intervention in the local slave economy.[13]

The Civil War and Postbellum Black Washington

The arrival of the Civil War presaged a dramatic expansion of civil and political rights for the district's black population, which had begun demanding equal treatment as the war began. The abolition of slavery came first—with compensation for their owners—in April 1862, the nation's first emancipation law. Local abolitionists celebrated in front of the White House, where President Lincoln gave them "a modest bow," and the event is still commemorated to this day. Laws establishing public education for African American children, eliminating the city's black codes, and ending other discriminatory statutes quickly followed.[14]

Because it bordered the Confederacy, Washington became a natural refuge for blacks who escaped from bondage during the Civil War. Union troops returned to the capital with former slaves they had confiscated, and some of the black soldiers who fought for the North settled in Washington after the war, joining wives, sisters, and daughters who had already moved to D.C.[15] As a result, the black population of the city tripled between 1860 and 1870, from 11,000 to over 35,000, expanding to become nearly one third of Washington's entire population (see figure 8.1). Nearly half of the newcomers were refugees from slavery; many were jobless, largely destitute, and uneducated, straining the federal and local government's ability to care for them.[16] Heeding the pleas of black Washingtonians and ignoring the city council's insistence that "the white man, being the superior race, must . . . rule the black," Congress gave blacks in D.C. the right to vote in local elections in 1867, three years before passage of the Fifteenth Amendment granted the suffrage to African Americans nationally. Two years later, Congress had ended all references to race in the charters of Washington and Georgetown, and Washington City—now under more racially tolerant leadership—passed laws in 1869 and 1870 banning racial discrimination in various public establishments.[17]

The late 1860s and early 1870s were a time of great activity and promise for Washington's African Americans. Blacks quickly emerged as an important political force in the city, voting in high numbers, actively participating in Republican Party affairs, and electing several African Americans to local office.[18] They made what one scholar called "upstart claims," sometimes successfully, for such rights as the freedom to use "white" streetcars and trains and visit the galleries of Congress. New schools for African Americans were authorized: Howard University, chartered by Congress in 1867; Preparatory High School for Colored Youth, the nation's first public black high school, in 1870; and the Sumner School in 1872, housed in a building designed by the prominent architect Adolf Cluss (and still standing today). Washington's growing black middle class included African American doctors, store owners, government workers, and teachers. A handful, such as James Wormley, who built a famous (and mostly white-only) hotel in 1871, became quite wealthy.[19]

Nonetheless, racism and the difficulty of integrating so many former slaves into the local economy were daunting challenges. Many African Americans were "confined to the most menial and lowest paying jobs" or hindered by white labor unions that instructed its members not to work alongside blacks. Some neighborhoods, such as LeDroit Park (near Howard University), went from mostly white to mostly black, but others became or remained strictly off-limits to African Americans. This artificial restriction on affordable housing contributed to the growth of black ghettos along the outer edges of the city and in Washington's crowded, unsanitary, and neglected alleyways, where African Americans would be out of sight yet "close enough to do the 'dirty work' traditionally left for newcomers," as one historian put it.[20]

Even worse, with few white Washingtonians enamored with the idea of black suffrage or civil equality, the political and social rights of African Americans in the district depended on the unreliable protection of Congress. Not many G.O.P. congressmen had ever been fully committed to complete racial equality—certainly not as an enduring cause—and many grew weary of upstart claims, especially the push by some for integrated city schools. A strike by black laborers in mid-1871 and the budgetary problems of the (pro-black) government led many members of Congress to question whether African Americans in Washington deserved equal rights at all. In 1874, Senator Oliver Morton (R-IN) warned that it would be "a step backward" for Congress to listen to the "many people in the district who are willing to be disfranchised for the purpose of getting clear of the colored vote," but a Republican-led Congress nonetheless voted that year to suspend all district government. Washington earned the dubious distinction of the first place in postwar America to disenfranchise blacks, serving as a model for Southern states that would do the same in the coming decades. When the largely Southern Democratic Party regained control of the House of Representatives months later, the fate of Washington's black community was sealed. It was left without political recourse against discrimination, segregation, and racism.[21]

However, the city's black population found ways to cope, even thrive. For one thing, the federal and local government still provided many African

Americans in Washington with stable, skilled employment. One of the most prominent such jobs was the city recorder of deeds, responsible for keeping track of property records; the first black recorder of deeds, appointed in 1881, was Frederick Douglass, the famous former slave who by then had established himself as a major figure in the city's black community. African Americans also developed a "secret city" of economic, educational, and social institutions in the last decades of the nineteenth century.[22] Black churches, long an important basis of communal activity and support, remained so. For instance, the African Methodist Episcopal (A.M.E.) Church, which dated back to the 1830s, created the Bethel Literary and Historical Association in 1881, a group that hosted talks by some of the most influential African Americans in the nation, including Frederick Douglass and Booker T. Washington, for over three decades. New periodicals reported on happenings in the black community, and an emerging African American elite founded secret societies and social clubs. Public schools for African Americans "became the best of their kind in the country," while Howard University overcame assaults by congressional Southerners to emerge as the nation's flagship university for African Americans.[23] Given Washington, D.C.'s political prominence and the size of its black population, establishing and strengthening such institutions was not only possible but, in retrospect, seemed almost inevitable.

Well-paying jobs (albeit lower paying than equivalent work for whites) and strong community institutions meant "greater opportunities for economic advancement for African Americans [in D.C.] throughout the Progressive Era than any other city in the South." These opportunities in turn attracted more African American migrants. By 1900, Washington's black community was the biggest of any city in the nation, and its elites became "an example for black communities everywhere," routinely discussed in African American newspapers and journals. Though its upper echelons garnered the most attention—and its members often distanced themselves from poorer African Americans in the hopes of becoming the first to be assimilated into white society—less elite blacks in Washington had their own organizations and churches too, and even impoverished alley dwellers developed "kinship networks" of support.[24]

Growing Racism and New Migration

Pressures on black Washingtonians increased in the early 1900s. Across the country, rising nationalism and an emphasis on unity—particularly on racial terms—became the foundation of a new "Rooseveltian nation" in which African Americans were ignored or marginalized. Presidents Theodore Roosevelt (1901–1909) and William Howard Taft (1909–1913) were among many white Republican politicians who favored the "go slow" approach to civil rights advocated by prominent black educator Booker T. Washington. Both presidents gave Mr. Washington exclusive authority over patronage jobs for African Americans, and he shut out long-standing members of the city's black community with whom he had no close ties. Segregation began to creep into the federal government. Some agencies, including the Treasury Department, started separating its black and white workers; fewer government jobs were made available to African Americans; and President Taft proved willing to refrain from hiring blacks for certain positions if a white person objected.[25]

Racism in the city remained powerful and pervasive. Southern lawmakers excoriated Roosevelt after he invited Washington for dinner in 1901. References to the capital's antidiscrimination laws of the 1870s were omitted when the city redrafted its statutes, a move that "encouraged white men to ignore them."[26] Mary Church Terrell, a prominent black civil rights leader, bluntly explained life for the city's African Americans in a 1906 speech. "Washington, D.C., has been called 'The Colored Man's Paradise,'" she observed, but "it is certain that it would be difficult to find a worse misnomer for Washington." She continued:

> The colored man alone is thrust out of the hotels of the national capital like a leper. As a colored woman I may walk from the Capitol to the White House, ravenously hungry and abundantly supplied with money with which to purchase a meal, without finding a single restaurant in which I would be permitted to take a morsel of food, if it was patronized by white people, unless I were willing to sit behind a screen.[27]

Things would get even worse for black Washingtonians after Woodrow Wilson was elected president in 1912. More government jobs were denied to blacks, and some African Americans were fired outright and replaced by white workers. What opportunities there had been for promotion in the federal service vanished. Allies of Booker T. Washington found, to their dismay, that he was no longer consulted for patronage appointments. Positions traditionally held by African Americans, such as the city's recorder of deeds and the register of the Treasury Department, were filled by whites. The federal government was also segregated to a greater extent than ever before: bathrooms and eating areas were designated as "white" or "colored," and screens were put up around one black Navy employee's desk so whites would not have to see him. Congressional Democrats tried to codify such segregation in the government and extend it to the entire city by introducing bills to separate streetcar seating by race and ban all interracial marriages. Although the bills failed, racial separation and discrimination spread informally, even into institutions that had heretofore avoided it. Washington's Catholic University, which Mary Church Terrell had once praised for its nonracial admission policy, began denying admission to blacks in 1914 and did not readmit them until 1936.[28]

By making their life increasingly hard, Wilson and congressional Democrats inadvertently galvanized black Washingtonians. The nascent D.C. office of the NAACP soon "grew into one of the largest branches in the country," and it protested the racist bills of Congress and a Wilson White House that, in the words of three NAACP officials, "has set the colored apart as if mere contact with them were contamination." Black city leaders also took advantage of their location in the seat of government to advocate for the civil rights of all blacks. In 1915, for example, local ministers lobbied against a proposal to ban blacks from immigrating to the United States. In the summer of 1919, worsening race relations led to an outbreak of violence as white mobs attacked African Americans in over twenty cities, including the capital. But armed black D.C. militias fought back, and the city's African American community became newly empowered and unified.[29]

Thousands of Southern blacks continued to migrate to D.C., particularly from Virginia and the Carolinas—part of the Great Migration of African Americans in the early decades of the twentieth century.[30] Many found jobs and support networks among black churches and social clubs organized around their state of origin. But while welcomed by some black Washingtonians, others greeted them with disdain and mistrust. In fact, skin tone and place of birth—often in combination—had long been a source of potential tension within the African American community, with favor given to those of fairer skin and a Washington freedman heritage. The black poet Langston Hughes was one of many who derided this hierarchy. In his memoirs he described "the 'better class' Washington colored people [of the 1920s], as they called themselves . . . as unbearable and snobbish a group of people as I have ever come in contact anywhere," whereas the city's "ordinary Negros" on Seventh Street offered "sweet relief" from such "pretentiousness."[31]

"Jim Crow was there" in D.C., one black immigrant put it, but "it was still not the South to us." Racial segregation remained a degrading fact of life, however. One woman who had moved from North Carolina remembered how "because you were black you could stand on the bridge, but not sit on the steps to listen to concerts or watch fireworks near the Capitol." Recalled another who was born in D.C. in the 1920s, "We couldn't go into Garfinckel's [a prominent department store]—not even black maids or mulattos. No, no. It made us feel very sad."[32] Also, while some places, including parts of Georgetown and the city's southwest quadrant, were racially mixed in the early twentieth century, most neighborhoods were not.[33] Real estate agents routinely refused to show properties in white neighborhoods to blacks, and banks would often not give mortgages to buy them. In 1912, one African American home buyer on an almost entirely white block of S Street, N.W., was "not permitted to see inside the house he was buying," recalled his daughter, and almost all of the block's white residents moved away within a decade.[34] In response to such incursions, white neighborhoods increasingly adopted racial covenants, which were agreements among homeowners not to sell or rent their property to blacks (and, often, Jews). Many congressmen stood solidly behind segregation in the capital—particularly Southern Democrats who often chaired the committees with jurisdiction over the district. Senator Pat Harrison (D-MS) lamented on the floor of Congress in 1926 that "the usurpation of white sections [of D.C.] by the colored people is destroying [home] values and shifting residential sections . . . [whereas] restricted areas for whites and colored make for the stability of [home] values and the common contentment of both."[35]

Despite Senator Harrison's fears of "usurpation," it was very difficult for African Americans to live outside of certain areas. The "undisputed heart" of the African American community since around 1900 was on U Street, N.W., in the LeDroit Park and Shaw neighborhoods. The area was home to a diversity of African American churches, hotels, banks, and other businesses and boasted a vibrant music and theater scene that earned U Street the moniker "Black Broadway." Washington native Duke Ellington played there, and dozens of famous black singers and musicians, including Louis Armstrong, Sarah Vaughan, and Ella Fitzgerald performed at the neighborhood's famous Howard Theater.[36]

Black Washington from the 1930s to the 1960s

The Great Depression hit the black community hard; many African Americans lost their jobs and were pushed to the city's economic margins. Though the federal government remained an important employer for African Americans, its efforts to rejuvenate the national economy often did little to improve—and occasionally even worsened—the lives of Washington's black residents. Most notably, the Federal Housing Administration (FHA), created in 1934 to save the nation's ailing housing market, issued regulations that encouraged suburban development over inner-city improvements and, in the words of one FHA manual, kept "inharmonious racial groups" from living together. As a consequence the urban exodus to the suburbs "was almost exclusively white," and African Americans in Washington and other cities saw little betterment of their own neighborhoods. Racial segregation also remained pervasive in the 1930s and 1940s, giving the capital a distinct Southern aura. Upon visiting the city in 1948, a German journalist observed that "I saw no

Negroes among the guests in any of Washington's hotels or restaurants, but all the more among the employees as maîtres d'hôtel, waiters, maids, doormen, shoeshiners, etc. *That is the Southern solution*."[37]

Yet a number of Washingtonians, black and white both, fought to end Jim Crow, and with gradual success. In the 1930s, local African American leader John Aubrey Davis formed the New Negro Alliance, a pioneering group that used both negotiations and boycotts to get many D.C. restaurants, grocery stores, department stores, and shoe shops to hire African Americans. In 1947 local actors boycotted the city's National Theater because it would not admit blacks, and in December of that year black parents refused to send their children to school because of the city's overcrowded and underfunded segregated educational facilities.[38] Meanwhile, federal urban policies continued to do more harm than good for black D.C. neighborhoods. In 1945, Congress created an agency to develop "blighted" areas of Washington. The result was "the largest urban redevelopment project undertaken in the United States" at the time. Trading social justice for beauty (in the words of Howard Gillette, Jr.), the government systematically destroyed the city's largely black southwestern quadrant, forcing its poorer residents into substandard housing projects and replacing their homes with architecturally sterile (albeit then-fashionable) apartments.[39]

Civil rights activism continued in the 1950s and early 1960s. In 1950, four local civil rights leaders (including Mary Church Terrell, then in her eighties) asked to be served at Thompson's Restaurant in downtown Washington. When the manager informed them that "we don't served colored," the restaurant was taken to court for violating the city's long-ignored anti-segregation laws of the 1870s. In a landmark 1953 decision, the Supreme Court agreed that segregation was illegal in D.C.—a decade before Congress banned segregation nationwide. The efforts of Terrell and many other individuals, groups, and institutions in the city to end segregation—together with the symbolic status of the capital's black community and the growing power of the African American vote in the North—led national politicians to take more seriously the plight of blacks in the district.[40]

Some of the steps that nationally elected leaders took were merely symbolic or declarative. When the Daughters of the American Republic would not let black opera singer Marian Anderson perform in their Washington theater in 1939, First Lady Eleanor Roosevelt and other prominent Democratic leaders sponsored a free concert for Anderson at the Lincoln Memorial. In 1940, the Democratic Party added a sentence endorsing D.C. voting rights to its party platform. President Harry Truman established a commission six years later to recommend ways to improve civil rights; its 1947 report, issued shortly after Truman became the first president to speak before the NAACP, called for an end to the "shamefulness and absurdity of Washington's treatment of Negro Americans." In 1948, the National Committee on Segregation in the Nation's Capital (which included Mrs. Roosevelt as a member) issued a damning report on racism and discrimination in D.C.[41]

But the federal government sometimes took more substantive action too. Roosevelt issued an executive order banning discrimination by defense companies doing business with the U.S. government. In 1948, Truman ordered an end to racial discrimination in all government agencies, and his administration raised the specter of a federal takeover of city swimming pools, parks, golf courses, and other sports facilities if local officials did not desegregate them. President Eisenhower continued in Truman's footsteps, appointing a new D.C. commissioner, Samuel Spencer, who worked to end discrimination in local government and city restaurants. After the Supreme Court ruled that segregated schools were unconstitutional in 1954, Eisenhower urged Washington to be "a model for the nation" by integrating its schools quickly.[42]

The black population of many American cities was growing during this period, but the pace of that growth in Washington was remarkable. Between 1950 and 1960, its African American population grew by nearly 50 percent, and, as noted in the previous chapter, Washington became an all-black city—the first such city outside the South.[43] Meanwhile, many whites in Washington, as in other cities, resisted desegregation, even after the Supreme Court ruled in 1948 that racial covenants were no longer legally enforceable. When their resistance failed, they voted with their feet: in the 1950s the number of whites in D.C. fell for the first time in its history, and by a steep 33 percent (see figure 8.1). Cynical real estate agents fueled the emigration by scaring residents of all-white neighborhoods into selling their homes cheaply before blacks might move in, then reselling those houses to African Americans at a huge markup. Though certain neighborhoods—like Shepherd Park, at the northern tip of the district—fought against this practice, known as blockbusting, many others succumbed.[44]

With Washington now a majority black city, its long battle for voting rights became more directly connected to the cause of civil rights, which the Democratic Party came to embrace by the mid-1960s. A constitutional amendment (which, tellingly, was ratified by only one state from the old Confederacy, Tennessee) gave the district the power to cast ballots for president starting in 1964. Washingtonians were granted the right to vote for school board in 1968, a nonvoting delegate in the House of Representatives in 1970, and its own local government in 1973.[45]

The 1970s to Today: New Challenges and New Opportunities

Politically and socially, by the mid-1970s African Americans had more opportunities in Washington and the nation than ever before. Blacks were elected to local public office; homes in all-white middle-class neighborhoods were available to would-be black homeowners; and new black-owned businesses emerged and thrived. Government employment in D.C. was, as always, a critical source for work; in 1980, nearly half of African American Washingtonians were employed by the federal or local government.[46]

African American culture, long an important influence, continued to make an indelible imprint on city life, shaped in turn by the "Carolina culture" that recent arrivals from the South brought with them. Local artists and performers like Chuck Brown, a rock guitarist who introduced an upbeat form of funk music called "go-go" in the 1970s, became nationally renowned. The city was proudly touted as a "chocolate city" where the ideals of the African American community could come to fruition.[47]

But the end of segregation, the flight of investment from cities, and the continued arrival of African Americans to D.C. also brought new challenges. Many black newcomers suffered from poverty and a lack of education, and not enough were able to land the kinds of blue-collar manufacturing jobs that had been the traditional source of employment for African Americans. Without the (admittedly forced) concentration of black residents, neighborhoods like U Street began to fall into decline. In April 1968, instigated by the assassination of Martin Luther King but more generally angry at their lack of economic opportunities, young blacks rioted in cities all over the country; in Washington, they caused massive property damage in the black commercial districts of U, Seventh and Fourteenth Streets, and much of it remained unrepaired for decades. African Americans with means moved to the suburbs of Maryland and Virginia, contributing to a decline of the city's black population after 1970. As we have noted previously, "white flight" became "wealth flight."[48]

Poor black ghettos emerged in Washington, as in other cities, where residents with few good job prospects lived without hope in what D.C. sociologist Elliot Liebow called a "sea of want." Some turned to the drug trade to earn needed revenue, further contributing to neighborhood decline. In the 1980s, the influx of cocaine exacerbated the problems of crime and drug addiction and, as the competition for clients by drug dealers and gangs became violent, rates of homicide—almost entirely of African Americans—went through the roof. The city became known as a murder capital as well as a political one.[49]

Although these problems were extraordinarily difficult to solve, local leaders did make efforts to address them. "Mayor for Life" Marion Barry secured private- and public-sector jobs for local blacks, particularly the young, and steered government contracts toward African American companies. The city built a new government edifice, the Reeves Municipal Center, on Fourteenth and U Streets in 1986, the epicenter of the 1968 riots, to encourage the revitalization of the neighborhood, which also benefited from a new Metro station in 1991. Murder rates began to decline from their 1991 peak of 482 murders, and in 2012 they had fallen to just 82, the lowest level seen since 1960.[50] In the D.C. suburbs, which has had the largest—and one of the nation's wealthiest—black suburban population in the country for several decades, the picture has been even brighter for African Americans.[51] Much of this has been due to the job opportunities provided by the federal government to black suburbanites. But it has also been a statistical artifact of the city's artificially fixed boundaries: "In most other cities," wrote two scholars, nearby black suburbs such as those in Prince George's County "would have been annexed [by Washington] long ago."[52]

Today, Washington and its larger metropolitan area remain a testament to the influence and importance of its African American residents. Almost all of the city's elected leaders are black; the region is home to hundreds of successful black businesspeople and entrepreneurs; and it remains a mecca for black music, arts, and scholarship. What would have once been unbelievable is now on the verge of reality: a new African American museum on the National Mall, in the shadow of the Washington Monument and mere blocks from a former slave auction site. Although the city's black population has been in decline, Washington remains attractive to newcomers of all ethnic backgrounds, including black gentrifiers.[53] Given all of this, the recent drop in the number of black residents seems unlikely to change the city's identity as a lodestar for African Americans around the country.

Washington as a City of Immigrants

Another dimension of Washington, D.C.'s ethnic legacy can be found in Judiciary Square, an area just east of the city's bustling Chinatown neighborhood. Judiciary Square is dominated by condo complexes, courthouses, and big government offices. But sitting hidden behind one of these nondescript buildings is the Holy Rosary Catholic Church, a small white edifice from the 1920s, which holds services in Italian as well as English. The former Adas Israel Synagogue is tucked away just a block north, moved from its old location on Sixth and G Streets, N.W., in the 1960s to make way for an office building. And on Fifth Street between G and H towers the gray stone St. Mary's Catholic Church, the center of religious life for Washington's German community in the nineteenth century.[54]

These three houses of worship serve as reminders of what the area once was: the home of several significant European immigrant communities. This may seem odd, given that Washington, D.C., is rarely thought of as a city of foreign immigrants. In a 2002 *New York Times* article that compared Washington unfavorably to New York City, Frank Rich wrote that by the mid-twentieth century D.C. "had long since missed out on the great wave of turn-of-the-century immigration" that provided "human and cultural variety" to other urban centers and that "even now, the capital lacks the ethnic spectrum of other major American cities."[55]

But Rich's characterization of Washington as an ethnically impoverished city is, at best, highly misleading. True, the capital's long-time status as a city of politics, not manufacturing, has meant that would-be immigrants have found less of the labor-intensive work traditionally open to new overseas arrivals. But this did not mean that the foreign-born avoided the city. The economic stability of a "government town" and the occasional local public works projects undertaken by the national government have long made Washington attractive to waves of immigrants seeking construction and service jobs. Some have arrived first in other American cities, then later moved to the District of Columbia for more desirable work. They have frequently brought the rich cultural practices of their homelands to Washington and created self-supporting communities, occasionally taking advantage of their location in the nation's capital to lobby for aid for themselves or for fellow immigrants around the country. Nor is this purely a historical phenomenon: in recent years, the larger metropolitan region has been one of the most popular immigrant destinations in the entire country.

Several noteworthy immigrant groups have called the district or its surrounding suburbs home over the course of history. These include Europeans, particularly Germans, Italians, and the Irish; Asians, notably the Chinese; Central Americans; and Africans, especially Ethiopians. We discuss each briefly in turn.

European Immigrants in Washington

It did not take long for foreigners to arrive in Washington. Shortly after the city was founded, men were recruited from France, Germany, Ireland, and Scotland to help build up the new capital. Many chose to settle in the emerging city, in places like southwestern D.C.; the Judiciary Square area, which was preferred by Germans; and Swampoodle, a neighborhood north of Capitol Hill that was favored by the Irish and became "notorious for its overcrowding and violence." As the city continued to grow, more Europeans arrived. Famine in Ireland drove many Irish to Washington; by 1850 they made up almost half of the city's foreign-born population. They were not always welcomed. Irish immigrants especially faced discrimination and open hostility. A candidate from the city's anti-immigrant, anti-Catholic Know-Nothing Party was elected mayor in 1854, and an organized effort to keep the Irish from voting in the city's 1857 elections led to the deployment of Marines to tamp down on street violence.[56]

Violence and discrimination did not deter Europeans hoping to find a better life in the nation's capital. A number of them became noteworthy members of the Washington community. Christian Heurich, who was born in Germany and moved to Washington in 1866, founded an enormously successful brewery and built for himself an imposing, fireproof brownstone near Dupont Circle that still stands. Jewish immigrants, mostly from Germany, established the Adas Israel Synagogue in Judiciary Square in 1876, and they were joined by Jews from Russia and Eastern Europe in the late nineteenth and early twentieth centuries. Italians also arrived in increasing numbers, lending their stone carving and construction skills to the creation of such city landmarks as the Library of Congress's Jefferson Building, Union Station, the Federal Triangle complex, and the National Cathedral. Some lived in Swampoodle, others in Judiciary Square, where Father Nicholas DeCarlo, a biology student at Catholic University, founded the Holy Rosary Church.[57]

Places of worship served as critical centers of community life for Washington's European immigrants. Émigrés also developed institutions designed to help their neediest members. For instance, D.C. Germans established an asylum for orphans in 1879, and in 1915 a Hebrew Home for the Aged was created to help elderly Jewish immigrants. Nor did they shy from using their prominence as Washingtonians to garner help from ethnic brethren in other American cities. In the mid-1920s, for example, Washington Jews solicited donations from Jewish Americans around the country, especially "national Jewish leaders [who] wanted the symbol of a strong Jewish presence in the capital," to construct a new community center on the prominent avenue of Sixteenth Street.[58]

Congress's absolute authority over the district meant that, unlike in other cities, the federal government could have a significant effect on the day-to-day lives of European immigrants in Washington. Foreigners were banned from owning property in D.C. (or in any American federal territory) in 1885, and Congress sometimes "intervened in deciding who could declare that meat sold in Washington butcher shops was kosher." When Congress put tight restrictions on the sale of alcohol in the district shortly before national Prohibition was enacted, it curtailed the social and economic life of Germans and other immigrants that revolved around the consumption of beer and wine. But this did not stop European Americans from using their proximity to power to lobby on behalf of national issues of import. For instance, Washington Jewish organizations submitted a brief on behalf of the civil rights lawsuit involving Thompson's Restaurant, and the D.C. Jewish Community Council helped organize local participation in the 1963 March for Jobs and Freedom.[59]

As in other cities around the country, European communities in Washington's urban core began to decline in the early to mid-twentieth century. Immigration restrictions in the 1920s sharply curtailed the inflow of European emigrants. Those who were in the central city gradually moved to outer neighborhoods in the district, or to Maryland or Virginia, and their departure accelerated following World War II and desegregation—part of the city's "white flight." Though far fewer foreign-born European immigrants live in Washington today, the descendants of those who first came many decades ago remain part of the greater Washington metropolitan area. Greek restaurants founded by Greek-born citizens can be found in Virginia and Maryland, and the Jewish Community Center is in Rockville, Maryland. And, of course, the many edifices their ancestors helped to build in the capital still stand today.[60]

The Chinese and Other Asians

It takes considerably less searching to find evidence of Washington's Asian community; the capital's Chinatown neighborhood is lively and decorated with many colorful signs in Chinese script. But it is not the original Chinatown of D.C.—nor, many would argue, the most authentic. And, in fact, immigrants from many other Asian nations also call the Washington region home, including Koreans, Vietnamese, and Cambodians.[61]

The Chinese were among the earliest, and largest, Asian nationalities to establish a foothold in the capital. The first immigrants arrived in the 1850s, drawn by job prospects and, in some cases, to escape discrimination and racial violence on the West Coast. Despite anti-Chinese prejudice, the community slowly grew, and Chinese grocery stores, import companies, and laundries sprouted up along Pennsylvania Avenue, near what is now the National Gallery. Like immigrants from Europe, the Chinese established their own community institutions, most notably "tongs" that offered "protective, charitable, and governing functions, thereby easing the immigrants' and migrants' transitions."[62]

In the 1920s, the U.S. government decided to construct a new building complex—today's Federal Triangle—along Pennsylvania Avenue. To save their community, the Chinese relocated several blocks north. But they had to overcome resistance to do it: residents of the area tried to get the U.S. government to block the move and even bought up properties to keep them away from Chinese buyers. In what seems now like a remarkably misplaced fear, whites who lived in the neighborhood worried that the new Chinese residents would ruin real estate prices and "not attract new business to the area."[63]

The neighborhood did, in fact, struggle for many years, especially when the Chinese, like other urban immigrant groups, began moving to the suburbs after the Second World War.[64] In the 1990s, however, the area was targeted for a slew of new developments: an indoor sports arena, a movie theater, and a bowling alley, plus apartments, chain restaurants, and stores. Community leaders insisted that all commercial signage be written in Chinese as well as English, and many new buildings included architectural details to evoke the Orient. But whether the area can still be considered a true ethnic neighborhood is debatable. Chinese Americans do live in the area, and, given how modernized China itself has become, D.C.'s Chinatown may look more authentic than one might think. Yet others have criticized the neighborhood's "Disneyfication," in which a chain restaurant such as Hooters has a veneer of exoticism simply by dint of a sign in Mandarin that reads "Owl Restaurant."[65]

What is true is that the Asian population in the Washington area has become both more diverse and more dispersed. Asians made up the second largest set of D.C. metropolitan residents born outside the United States in 2000, but no single nationality predominated among them. One of the area's biggest collections of Vietnamese stores is not in the city of Washington but in the Eden Center shopping mall, located seven miles away from D.C. in Falls Church, Virginia (see figure 8.2). Rockville, Maryland, has become a

Figure 8.2 Entrance to Eden Center Shopping Mall, Falls Church, Virginia
Photo by Matthew Green

Chinese "sociocommerscape," an economic and social hub catering to a specific immigrant group. "It's the new Chinatown," declared one Rockville tea shop owner. It is a trend hardly unique to Washington, as Asians in cities around the United States have moved to the suburbs, and growing prosperity in China means many Chinese are moving back to their country of birth or not moving to the United States at all.[66]

Newer Immigrants: Latin Americans and Africans

In contrast to European and Chinese immigrants, who came to Washington in the nineteenth century, Latinos and African-born immigrants did not become a significant presence in D.C. until well into the twentieth century. Latinos arrived first. Puerto Ricans and Mexicans moved to the city in the 1930s and 1940s seeking government-related work. Employees of embassies from Spanish-speaking countries became the basis for a small but important Latino community in the Adams Morgan and Mt. Pleasant neighborhoods. They were joined by Dominicans, political refugees from Cuba, and others arriving via "chain migration"—migrants coming to places where their friends and family members have already settled. By the early 1970s, a former Adams Morgan church had become a focal point for community activity and entertainment, and the neighborhood had started a Latino festival, Fiesta DC, that is still held annually. A civil war in El Salvador led thousands to flee to the district in the 1980s, and even today Salvadorans make up the single largest percent of foreign-born residents of the Washington metro area.[67]

Latino immigration created new stresses. Conflict occasionally emerged between groups of different North, Central, and South American nationalities who saw each other as competitors for jobs and services. Latino immigrants also grew resentful of government officials who either neglected them or assumed they were in D.C. illegally. "There was built up, pent up frustration for many years of feeling marginalized," as then-Mayor Sharon Pratt-Kelly later put it, that finally broke out in violence: a three-day riot in May 1991, precipitated when the police shot a drunken Salvadoran man in Mt. Pleasant. The uprising led the D.C. government to dedicate more attention to the Latino community, which in turn "focused on establishing a Latino presence and place—a 'voice'—within the larger polity."[68]

Today, as in a number of U.S. metropolitan areas, Latinos are the largest proportion of the immigrant population in the capital region. They differ from Latinos elsewhere, however, in their household income, which is the highest of any metro Latino population; their high levels of education; and their higher rates of homeownership. Some have left the city for the suburbs, and many suburban Latinos skipped the city altogether. The unincorporated Maryland town of Wheaton, located less than four miles north of the D.C. border, provides a striking example of this ethnic suburbanization. In 1990, whites made up 60 percent of the town's population, but just two decades later they constituted barely a quarter of the total, whereas the percentage of Latinos jumped from 13 percent to over 40 percent. While this dramatic change has created some tensions in the suburbs, many long-time Wheaton residents welcome their town's new cultural diversity. They boast about their annual outdoor Taste of Wheaton festival, which features booths selling tamales, Peruvian chicken, plantains, and pupusas, not to mention Italian, Chinese, and Thai food.[69]

Another source of exotic food is the Washington region's African immigrant community, which is of even more recent vintage. D.C. boasts the second-largest number of African immigrants of any city in the country,[70] and, while those immigrants come from many countries of the continent, including Nigeria, Sierra Leone, and Ghana, a considerably large proportion are Ethiopian-born. In fact, though estimates of its size vary, many consider the area to be home to the largest Ethiopian community in the United States.[71]

Many Ethiopians moved to the United States in the 1970s and 1980s to study at American universities and to flee the country's civil war. D.C. in particular was attractive to Ethiopians for several reasons: it was the home of the Ethiopian embassy; Howard University, the nation's preeminent African American school; and a large black native population. Many settled along the traditional African American area of U Street, and it is there one finds restaurants with such Ethiopian names as Dukem, Abiti, and Queen Makeda. An effort to formally designate the intersection of Ninth and U as "Little Ethiopia," however, met with resistance from the city's native-born black community, who feared the usurpation of the neighborhood's ethnic history.[72]

Ethiopians are by no means unified, nor do they all live on U Street. Political conflicts in their home country occasionally divide the community—to the point that some D.C. Ethiopian soccer players who felt a stronger allegiance to their home country's government formed a rival tournament in 2012 to compete with an existing Ethiopian tourney. Catering to its growing suburban Ethiopian population, Silver Spring, Maryland, began holding an annual Ethiopian cultural festival with food, music, and crafts in 2011. Nonetheless, despite their political differences and *heterolocalism*—the lack of close spatial proximity—the Ethiopians of the capital region find ways to stay connected, including a newspaper (*Zethiopia*) and a directory of businesses catering to their community.[73]

Conclusion: The Changing Demography of a Capital City

Washington is not an ethnically homogenous city. Nor can its population be stereotyped as "black" versus "white." Besides being far less ethnically or racially segregated than metropolitan areas like Detroit, Chicago, or Los Angeles, the D.C. area is recognized as a major "emerging immigrant gateway" for those searching for a refuge from their home countries. The region is also characterized by a remarkable number of "ethnoburbs," suburban neighborhoods with sizable ethnic and immigrant populations. Thanks to the development and zoning regulations of suburban communities like Montgomery County, Maryland; the ready availability of suburban jobs; and the artificial proximity of out-of-state suburbs to the district due to Washington's (constitutionally fixed) borders, the D.C. region has witnessed a large, thriving, and growing foreign-born population.[74]

Of course, Washington, D.C.'s demography is distinctive in other ways besides the race, ethnicity, and country of origin of its residents. Youth is another remarkable feature of the city's population. The 2010 census showed that the population of those aged twenty to thirty-four grew by more than 20 percent over the past decade, and between 2009 and 2012 the metropolitan region saw the single biggest gain in population in the United States among people between twenty-five and thirty-four years old. Younger people now make up over three-tenths of D.C.'s entire population, helping inject a new vitality and energy into the district. In the words of three *Washington Post* reporters: "A city once renowned as a mecca for workaholics is starting to be thought of as a place that's fun."[75]

Historically speaking, however, the city's long legacy as a mecca for African American culture, politics, and community is what differentiates the city from all others, and its growing and largely suburban population of Latinos, African, and Asian immigrants gives Washington and its surroundings a unique look and feel. It is also a population that is very much in flux. The city's ethnic mix may look very different even a mere ten years from now—a possibility that makes the capital region one of the most exciting places to live and watch.

CHAPTER 9

The Economic Life and Development of a Capital City

Metro riders who arrive at the NoMa-Gallaudet stop in northeast D.C. encounter an impressive display of recent and ongoing development. Cranes swing far above the skeletons of new buildings that rise from lots once empty and abandoned. The glass and cement headquarters of the Federal Bureau of Alcohol, Tobacco, and Firearms, built in 2008, sits just steps from the Metro station. Elsewhere stand freshly built office buildings, apartment complexes, restaurants, and stores, and a Hilton hotel that abuts the Metro tracks. Even the station itself is new, built less than a decade ago.

Meanwhile, in a residential neighborhood just a short distance away, a man seeks refuge from sewage pouring out of his toilet. He lives in a basement apartment in LeDroit Park, one of Washington's earliest residential communities, whose residents have long been ignored by the city. When row houses in the neighborhood began converting their basements into separate apartments decades before, the area's aging sewer system proved unable to handle the additional load, and backups (and flooding during heavy rains) began to plague the area. Residents' complaints went unheard for many years, and only recently, with the arrival of new, wealthier residents, did city officials begin to take note of the problem. But no solution has yet been found, and in the meantime the arrival of those newcomers has put still more stress on the old pipes of LeDroit Park.[1]

These disparate images from two neighboring parts of the city underscore several important themes of Washington, D.C.'s economy. The ongoing development of the NoMa area is a visible manifestation of the capital's growing wealth and strong attraction for developers, even during economic downturns. Much of that wealth and economic success is due to the presence of the federal government—a unique advantage of Washington and one that often serves to protect the city's economic well-being in times of recession—though it is also the consequence of active efforts by the national and local government and the private sector to encourage development. From the perspective of Washington's residents, that success is not always welcomed, and for some—like the unfortunate fellow in LeDroit Park—it may have unpleasant, if not unforeseen, consequences. Finally, development projects like those in NoMa are hardly unique to the city of Washington. In fact, there is an important distinction between the economic life of the city itself and that of its suburbs in Virginia and Maryland, with the latter often doing much better economically than the former—though the dividing line between prosperity and poverty also cuts through the city and surrounding states.

A Brief Economic History of Washington

"Throughout America's national history," write two preeminent scholars of city politics, "the most fundamental goal of its cities has generally been local economic growth."[2] This was no less true of the city of Washington. From the very beginning, city planners wanted to ensure that the nation's new capital would be economically vibrant, and they never assumed that headquartering the national government would be enough to bring jobs and commerce there. One of the major reasons George Washington sought to locate the city on the Potomac was his belief that the river could be turned into a major inland waterway that would not only politically connect the country's inland to its coastland but also facilitate commerce. Having caught "Potomac fever," Washington even invested in a company that planned to make the river an easy route to transport goods from the interior for sale and trade in the capital.[3] As noted in chapter 1, Pierre L'Enfant's geometric design of the city was also intended in part to attract development. The city's multiple, state-specific squares were supposed to lure members of Congress and other government officials from those states to live there. Stores and businesses would follow, leading to a series of "mini-towns" scattered throughout the city that would, in L'Enfant's vision, gradually interconnect—a sort of prototype keno capitalism, the type of development (often seen in postmodern cities) in which distant urban centers gradually merge together.[4] As noted in chapter 1, L'Enfant—recognizing the economic as well as artistic advantages of canals—included in his plan a canal intersecting the city that would further help move goods to and from Washington.[5]

L'Enfant's city was to evoke grandeur as well, but its artistic merits were less important to local leaders than the need to attract investors and settlers, and the protests of the French architect that public spaces should be built before lots were sold for development were ignored. The city's economic development proved far slower and more haphazard than hoped, however. A local economy did emerge: small shops, eating establishments, and public markets were founded that served the city's transient political actors as well as its growing population of permanent residents. But the Potomac was found to be largely unnavigable, never becoming the thriving commercial waterway that Washington had hoped for.[6] In lieu of the river, a new canal, the Chesapeake and Ohio, was built to connect the neighboring riverside town of Georgetown to the interior, and got as far as Cumberland, Maryland. Yet even before the canal was completed, railroads had become a faster and more extensive means of transportation, and the canal, subject to frequent flooding, usually lost money.[7] Georgetown became an even less desirable port when ships powered by steam became commonplace and needed a deeper harbor than Georgetown's.[8]

After the Civil War, the city's unofficial "mayor," Alexander Shepherd, undertook efforts to grade city streets and open spaces for development. It was sorely needed. One newspaper described Washington's streets in 1869 as ones "where when it's dry you could not see where you were going, and when it's wet you can't go." Wanting a city that would proudly represent a united nation and seeking to counter pressure to move the capital to a more attractive and thriving city (most notably St. Louis), President Ulysses Grant stood behind Shepherd's efforts and successfully pushed for the construction of additional government buildings.

As Washington improved, it attracted new wealth in the form of Gilded Age entrepreneurs who had amassed great fortunes and sought a place to settle for the winter months. Not all of the nouveau riche were outsiders. The German brewer Christian Heurich, mentioned in the previous chapter, eventually became the city's biggest employer outside of the federal government. Fearing fires—his brewery had been damaged by several—he built a state-of-the-art brownstone mansion in the Dupont Circle area out of steel and concrete. Until his death in 1945 at the ripe old age of 102, Heurich resided in the house with his third wife, Amelia, twenty-one years his junior.[9]

Economic development and internal improvements continued after Shepherd's rule ended in 1874. In the 1890s, for instance, Washington was covered by more asphalt than any other American city save Buffalo, New York. National politicians also saw an opportunity to get rich by investing in the capital. In the late 1800s, Sen. John Sherman (R-OH) made a large land purchase north of the city, developing the property into housing and in the process creating the new neighborhood of Columbia Heights. Secretary of State John Hay purchased contiguous lots on Connecticut Avenue and L Street in the early 1900s, then demolished the buildings on those lots and put up a huge apartment building called Stoneleigh Court.[10]

As the nineteenth century made way for the twentieth, Washington continued to grow in population, wealth, and economic activity, thanks in no small part to the steady expansion of the national government. Over 50,000 federal government jobs were created in the city during the 1910s, nearly 35 percent more than all the jobs that had existed in Washington at the decade's start (though many were created during, and disappeared after, World War I). City boosters began employing the word *capital* rather than *metropolis* to reflect the government's growth and make Washington sound more appealing.[11] Residential neighborhoods steadily gave way to commercial ones. In 1926, noted the local newspaper *Evening Star*, lower Connecticut Avenue—once a quiet area where the wealthy and famous lived—had quickly undergone "a change nothing short of astounding," populated by "a bustling crowd of business people, shoppers and others engaged in commerce, who arrive in the street cars, in fast moving automobiles and commercial trucks." Morris Cafritz, a Lithuanian émigré who went from running a local grocery store to becoming a huge real estate mogul, built apartments and hotels on K Street, N.W., and elsewhere in Washington from the 1920s until his death in 1964. The New Deal and World War II brought even more businesses and workers to the city and raised Washington's national and international profile to new heights, contributing further to its economic growth. In 1940, for instance, over 40 percent of Washington jobs were in the federal government, a percentage far higher than even what the city has today.[12]

As with so many American urban places, the decades after the war were better for the suburbs than for the city of Washington itself. New housing developments were constructed with federal aid in the city's southwest section, but they displaced thousands of low-income Washingtonians while doing little to improve the capital's economic or social vitality.[13] Meanwhile, government programs that encouraged suburban development, including generous home loan subsidies and aid for freeway construction, pulled people away from Washington as they did from cities around the country. Transportation and communication improvements made possible what one urban scholar called a "megalopolis," the connection of several cities across hundreds of miles, from D.C. to Boston.[14] Racial prejudice also led many wealthy and middle-class whites in the United States to flee to suburbs in the decades after World War II. As noted in earlier chapters, this was no less true of Washington, contributing to its landmark status as majority black in the 1950s. Prejudiced white Washingtonians were further convinced to depart after the U.S. Supreme Court issued rulings that declared housing and educational segregation unconstitutional. Problematic in particular for Washington was its fixed boundaries, which limited the potential for new development within its jurisdiction, especially in contrast to the abundant space available in neighboring Maryland and Virginia. As a result, between 1950 and 1970 the greater urban area more than doubled in size from 180 to over 500 square miles, while Washington itself remained fixed at 68 square miles.[15]

Washington was one of many American cities that entered an "urban crisis" in the 1960s and 1970s as conditions in urban centers declined drastically. Jobs disappeared, real estate prices fell, and city governments were left to deal with growing poverty and crime rates, governing what was becoming "a preserve for poor blacks and single mothers struggling to survive" amid a declining tax base.[16] City populations, already on the decline, continued to fall, often at a faster rate. The destructive 1968 riots that hit many cities, including Washington, following Martin Luther King's assassination further drove out wealthy and middle-class residents. In 1970 the nation's suburban population exceeded city populations for the first time in its history. A recession later that decade delivered yet another blow to Washington and other cities. "By 1975," writes one historian, "the old central cities appeared to be going down the drain." Under those circumstances, it proved politically easy and popular for President Ronald Reagan to impose drastic cuts in federal aid to cities in the 1980s, making matters for cities still worse.[17]

The urban crisis was admittedly direr for other American cities than the nation's capital, especially since Washington did not depend on declining sectors such as heavy manufacturing and industry for its survival. For instance, D.C.'s population fell by 16 percent between 1970 and 1980, whereas during the same decade Detroit's population dropped by 20 percent, Cleveland's by 24 percent, and St. Louis's by 27 percent. The steady employment opportunities of the federal government undoubtedly helped Washington, but out-of-district suburbs nonetheless got a bigger share of the federal largess. Huge increases in defense spending, particularly during in the 1980s, benefited northern Virginia enormously. Home to the Pentagon and the CIA, the area soon specialized in what one magazine dubbed "death" industries.[18]

But the city did gradually start improving economically. Washington became a "hot spot for gentrification," with neglected neighborhoods like Adams Morgan growing in popularity and gays and bohemians serving as urban pioneers in places like Dupont Circle.[19] The capital also benefited from two big economic booms in the 1970s and 1980s. First, government lobbying mushroomed from a small cottage industry into a major multimillion-dollar enterprise, eventually employing thousands of lobbyists in dozens of firms—a change that "helped make greater Washington one of the wealthiest regions in America." Second, the city saw a spike in real estate development in the 1980s, with over 160 million square feet of additional office space built in Washington, a huge amount relative to other cities. Prominent new buildings constructed in that decade include the mammoth Avalon apartment building in upper northwest, One Franklin Square on Thirteenth and K Streets (one of the largest buildings in the district) and the glassy, futuristic Tech World office building near Chinatown. The city did its best to encourage such development by keeping business taxes low and making it easier to secure building permits, and it tried to bring more outside visitors with the construction of a new convention center in 1983.[20]

Neither the lobbying nor the real estate industries did much to improve the economic conditions of the city's neediest residents, however, and the real estate boom soon came to an end. By the early 1990s, with a recession and rising crime in urban areas across the country, it seemed that Americans were ready to give up on cities altogether. But around the end of that decade, Washington became part of a new American urban renaissance.[21] Crime and urban poverty rates fell, and young people found city life more convenient and enjoyable than suburbia.[22] Between 2000 and 2010, the downtowns of Washington and nearby cities saw the second-largest percentage growth in residents of any metropolitan area in the United States. The D.C. area was also helped by additional defense dollars spent on homeland security during the George W. Bush presidency.[23] With more people came increased demand for new apartments, grocery stores, restaurants, and cultural facilities and events that in turn brought additional economic investment.

The cityscape of Washington changed, and continues to change, as part of this revitalization. Twenty years after it was built, the capital's outdated convention center was demolished and replaced in 2003 by a new, three-block-long facility, while the old center's ten-acre location was slated to become "CityCenterDC," a collection of stores, offices, condominiums, and public space. A stadium to house Washington's new major league baseball team completely transformed the Capitol Riverfront area in south D.C. Other neighborhoods, including Columbia Heights, Chinatown, Logan Circle, and U Street, saw a blossoming of new apartment buildings, eating establishments, storefronts, and movie theaters—usually with financial and regulatory assistance from the local and federal government. Not even the "Great Recession" of 2008–2009 halted this economic revitalization, which continues to this day.[24]

Government as Regional Economic Engine

The singular puzzle facing every American city is how to attract and maintain capital and investment. Each city competes with others around the country—and increasingly the world—for a limited amount of business and commerce. New businesses can locate in many different places, and existing businesses may be tempted by financial incentives to uproot from where they are. So compelling is this need to attract commerce that urban scholars Dennis Judd and Todd Swanstrom identify the "politics of growth" as one of the three defining elements of city politics in the United States.[25]

The tall construction cranes in NoMa and elsewhere that dot the D.C. cityscape suggest that Washington has been winning that competition. And if we look at the greater metropolitan region, leaving aside for the moment the city itself, the statistics are impressive. In 2010, the area represented the fifth-largest regional economy in the country and had the greatest number of fast-growing privately held companies. Two years previously, the region had the eleventh-largest gross domestic product of any urban area in the *world.* Its residents could boast of the highest median income in the United States—over $85,000 in 2010—and a larger percentage of its workers have bachelor or graduate degrees than any other American metropolitan region.[26] The social scientist Richard Florida ranked the area as tied for tenth most "economically powerful" on the globe, as measured by gross regional output, financial power, and number of patents awarded.[27]

What accounts for this success? Most assume that the answer is the presence of the national government. And there is much truth to that assumption. The federal government is a direct employer of over 370,000 people, ranging from political aides and bureaucrats to consultants, lawyers, administrative staff, and maintenance and service workers.[28] In addition, tens of thousands work for companies that provide goods and services to the federal government. The government spends more on equipment, research, and other goods and services in the region than in any other state, and more in the city of Washington itself than in all but six states. The Defense Department is a particularly generous source of funds in this regard. Federal military spending has formed the basis of a major defense industry in the region, which includes some of the largest employers of the area, including Lockheed Martin (which employed 23,000 people in 2010), Northrop Grumman (which employed 20,000 people) and Science Applications International Corporation (which employed over 17,000). In addition, the desire of private companies, other governments, nonprofit groups, and intellectuals to steer government policy generates yet more economic activity. In fact, the city's many advocacy groups, law firms, and nonprofits make D.C. a natural landing spot for former government officials seeking employment. By one account, over two hundred people who once served in Congress work as government lobbyists.[29]

All of the regional economic activity that derives from the operation of the federal government is a source of considerable prosperity. Some of the city's great philanthropists and eminent residents got their start in government, such as David Kreeger, a New Deal lawyer who became wealthy from the insurance business and gave lavishly to arts and universities in the metropolitan area. Government spending also serves as a giant cushion to protect the capital from the worst effects of economic slowdowns. During the Great Recession, for instance, unemployment in the Washington region rose from 2.8 percent in November 2007 to 6.3 percent two years later, while during the same period the national rate climbed from 4.7 percent to nearly 10 percent (see figure 9.1). This disparity is nothing new; the historian Constance Green observed that Washington did not suffer terribly during a recession in 1819–1821 "for government operations materially lessened business stagnation in the federal city." It should be noted, however, that government spending is not a perfect shield from larger economic forces. The Great Recession led many companies and state governments to cut back on travel expenditures to D.C., hurting Washington-area hotels in particular.[30]

The powerful economic influence of the federal government implies a passive relationship between it and the region's economy: that the mere presence of the government has been enough to bring jobs and commerce

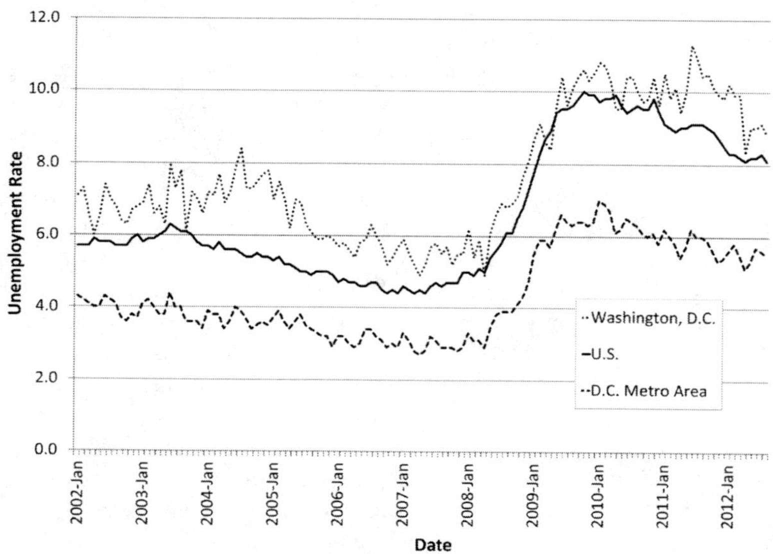

Figure 9.1 Unemployment Rates in the United States, Washington D.C., and D.C. Metro Area, January 2002–August 2012

Source: Bureau of Labor Statistics. U.S. data is seasonally adjusted; Washington and D.C. Metro area data is not.

to Washington. To be sure, Washington is but one of many cities that has an *agglomeration economy* in which particular businesses have found it economically advantageous and efficient to settle in the same place in order to be close to a skilled labor pool, similar industries, or lots of customers.[31] But just as George Washington, Pierre L'Enfant, and others sought ways to bring commerce to the new American capital, so too has active leadership by people both in and out of local and national government been necessary to usher in greater economic vitality and investment. After its creation in 1889, for instance, the Washington Board of Trade worked for decades to protect business interests in the city, create parks and new transportation routes to maximize local growth, and develop an economic strategy that tied D.C. to the Southern regional economy.[32] In the late nineteenth and early twentieth centuries, city leaders worked hard to promote the city as an ideal location for conventions and a place for patriotic citizens to visit its newest monuments and museums.[33] More recently, Virginia's and Maryland's state and local governments have done their best to make their areas attractive to government agencies, workers, and companies doing business with the national government.[34]

Washington itself, suffering from comparative disadvantages vis-à-vis its neighbors (as will be discussed in a moment), has been especially dependent on active leadership to bring new growth within its borders. The development of the NoMa neighborhood is a prime example of the critical role of city leadership and proactive public policy. Only when the city and national government, along with the private sector, contributed the necessary $120 million to build a new Metro subway stop was the potential for development in the area fully realized. In early 2007, the D.C. city council also created a Business Improvement District for the area, which allowed revenue from neighborhood businesses to be collected and spent on improving and encouraging more investment in NoMa.[35] Other thriving areas of the city, such as Columbia Heights, H Street, N.E., and the neighborhoods surrounding the new baseball park similarly depended on community leadership and the investments of public-private partnerships, often benefitting from the construction or upgrade of transit stations and tax breaks from local authorities. And the city's homeownership tax credit, enacted in 1997, created a major financial incentive for people to buy homes in the city versus the suburbs. According to one survey, it influenced the decision of 40 percent of all new home buyers in 1998 and 1999 to live in the city of Washington.[36]

Furthermore, despite the contribution of federal spending to D.C.'s economy, there are drawbacks to the government's large economic presence. Government buildings are tax-exempt, robbing the local government of needed revenue.[37] Companies that contract out with the feds can go under if their contracts are not renewed. If Congress fails to enact spending bills, shutting down the federal government—as happened in October 2013—the economic impact can be devastating for the city and the region.[38]

Even if the government is funded, cuts to the federal budget are an ever-present possibility. The county with the single biggest amount of procurement contracts in the nation—Fairfax County in northern Virginia—faced layoffs of 13 percent of its workers when the U.S. government began imposing major automatic spending cuts in early 2013. The two-week government shutdown in October 2013 cost the region an estimated $40 million in sales tax revenue alone. The city's long-neglected Anacostia neighborhood, which suffers from high unemployment and crime rates, hoped for a major influx of residents and commerce when it was designated to be the new home of the Department of Homeland Security in the mid-2000s. But construction of a four million-plus-square-foot facility to house the department was significantly delayed because of Congress's zeal to cut spending after 2010, leaving the neighborhood in limbo. These and other looming budget cuts led one local businessman to put the situation this way: "The goose that's been laying the golden egg—the federal government—has a bad case of pneumonia."[39]

Washington's government-fueled economic growth can also stoke the fires of resentment that many Americans have toward the city and harden their belief that D.C. is too distant from the country at large. The *New York Times* columnist Ross Douthat complained about a "gilded District" where "the wealth of Washington is ultimately extracted from taxpayers more than it is earned."[40] In a 2011 essay in the journal *Harper's*, author Thomas Frank argued that "Washington is indeed out of touch with the suffering of the nation" because its economic success "has persuaded its resident journalists and pundits and policymakers to credit all sorts of unsound economic ideas." Both Douthat and Frank ignore the poverty that lingers in the city, and Frank in particular assumes elected officials in Washington neglect their own constituents' economic conditions once they arrive in the nation's capital. But both do tap into a deeper vein of distrust Americans possess about their government that, as we noted in chapter 4, often manifests itself in suspicion about the city itself.[41]

Economic Diversity

The founders of the city expected that Washington would thrive and not solely, or even principally, because it was the center of national politics. To some extent, their vision of a city both economically strong and diverse proved correct. For as important as the federal government is to the area's economy, looking at Washington as a government town conceals considerable economic heterogeneity in both the capital and the greater D.C. area.

By one estimate, barely more than a third of the region's economy depends on the national government. Some 65 percent of city workers are employed in the private sector. Health care and education are especially important economic sectors; together they provide jobs for over 100,000 people in Washington itself and more than 360,000 in the metropolitan area.[42] In 2009, only one of the city's ten-largest local nonpublic employers was neither a university nor a hospital.[43] Education in particular has a long and storied history in Washington. George Washington, James Madison, and other Founding Fathers had hoped to establish a national university in the capital to train the nation's future leaders.[44] Though they failed, local priest John Carroll—later the country's first American Roman Catholic archbishop—founded a university in the city of Georgetown in 1789, and more institutions of higher learning were created in the late nineteenth and early twentieth centuries. Students came in increasing numbers to D.C. during the Second World War and again after the mid-1950s. The result has been an economic as well as intellectual boon for the nation's capital. In 2010, the fourteen universities located in the metropolitan area contributed over $5.6 billion in additional money to the region's economy, with 25 percent of that directed to the city of Washington.[45]

The region's nongovernmental economy includes more than universities or hospitals. "Private idea brokers"—lawyers, consultants, and public relations experts—populate the city in large numbers. Many can be considered part of what Richard Florida calls the "creative class"—those who are highly educated and innovative and who, according to Florida, not only bring greater cultural activity and diversity to a city but are themselves the foundation of strong economic growth.[46] Entrepreneurs have found new ways to make money in Washington, be it a gourmet food truck catering to the downtown lunch crowd or a Portland, Oregon, company opening a hugely popular bike-sharing service. Companies of all sort and sizes can be found in the D.C. area—companies such as MICROS Systems, a Columbia, Maryland, firm that supplies hardware and software to hotels, restaurants, and other retail businesses; Volkswagen, which moved its American headquarters to Herndon, Virginia, in the late 2000s; and Marriott, the hotel chain that got its start when two Washingtonians opened a chain of sweet shops in the city in the 1920s. Black Entertainment Television, founded in 1980 by Robert L. Johnson, an African American entrepreneur (and the nation's first black billionaire), has its headquarters in Washington. An international company based in Washington, LivingSocial, offers gift certificates for its 70 million members for a range of goods and services.[47] Companies with main offices elsewhere also serve as important regional employers, including Safeway and Wal-Mart.[48] In fact, the area's single-largest private-sector source of jobs—employing over 30,000 people in the area—is the McDonald's restaurant chain.[49]

The capital also has important economic ties to other countries. Though not an economically "global city" to the same extent as London, Paris, New York, or even Chicago or San Francisco, Washington nonetheless appeals to many countries seeking a safe and profitable place to invest in real estate and businesses. Oil-rich investors from the Middle Eastern nation of Qatar, for instance, invested $700 million dollars in the CityCenterDC development. In 2010, the area was also home to over a thousand foreign companies.[50] And, of course, Washington is a major tourist attraction for foreigners as well as Americans, who come spending their money in D.C. hotels and businesses.[51]

Certainly, a good amount of this private-sector activity is an indirect consequence of government spending. Well-paid federal employees buy new cars; car dealerships hire more workers; those workers buy houses and eat at restaurants; and so on. But to attribute the capital's entire economy—or even most of it—to government largesse would miss the many diverse businesses and economic activities that make Washington one of the most affluent and prosperous cities in the United States.

City vs. Suburbs, East vs. West, Rich vs. Poor, Black vs. White

The city of Washington may benefit from the presence of a large, wealthy, and powerful national government and a rich diversity of private-sector jobs. But it is not without economic challenges. These include its immutable borders, the strong incentives for businesses and government agencies to move to surrounding suburbs, significant economic inequality, and the detrimental consequences of economic growth.

The district's constitutionally fixed boundaries limit its ability to expand the amount of taxable land available for investment and development. As we noted in chapter 7, its geographic size has long been smaller than that of other metropolitan centers that managed to expand their borders in the late 1800s and early 1900s. To be fair, for decades many American cities have seen companies and factories move out of their boundaries and into neighboring communities and counties that refused to let themselves be incorporated into the metropolis. But the typical city can, at least in theory, regain some of that tax revenue from the state in which both it and its suburbs are situated. Not so D.C., which has no way of reclaiming the lost sales and real estate taxes from businesses and citizens that move to Maryland and Virginia.[52]

The states surrounding D.C. are also able to attract new businesses that Washington cannot. The Metro subway system allows easy transportation from the city to parts of northern Virginia and southern Maryland, and the Beltway—the freeway that encircles D.C.—permits easy mobility between both states. The city's airports are all located outside of the district. Maryland and Virginia offer cheaper real estate and lower taxes for businesses than can be found within Washington. They do so with favorable tax laws and less restrictive development rules, especially on the height of new buildings, which make the suburbs more attractive to developers and would-be downtown residents.[53] One need look no further for evidence of this last point than the skyscrapers of Rosslyn, Virginia, that abut the D.C. border—just across the Potomac River and only one Metro stop from the city—which are too tall to have been built legally in the city of Washington (see figure 9.2).

As a consequence, the city proper has seen less economic success than the larger metropolitan area. According to one study by George Mason economist Stephen Fuller, 60 percent of all new jobs created in the region between 1980 and 2010 were in northern Virginia, not Washington. Unemployment rates within the district are always far higher than in greater D.C., and almost always higher than the national average (see figure 9.1). Meanwhile, centers of economic development and activity have emerged, or are emerging, in places like Tysons Corner, where leading-edge technology companies have

Figure 9.2 The Rosslyn, Virginia, Skyline
Photo by Matthew Green

made the area an "Internet Alley"; and Gaithersburg, Maryland, the location of the "Life Sciences Center," an expanding collection of biotech companies, university centers, and a hospital.[54] More recently, Tysons Corner has renamed itself "Tysons" and hopes to use a new Metro line to convert its car-oriented shopping areas into pedestrian-friendly mixed-use communities.

This does not mean that all of the D.C. suburbs are better off than the city proper. Some job growth has returned to the capital in recent years. One study from the 1990s found an economic dividing line not between Washington and its suburbs but along a north-south axis cutting directly through the district, with the eastern side of the axis "bear[ing] the burden of poverty." With some exceptions that line still exists today. In February 2013, Maryland's Charles and Prince George's counties, which lie east of D.C., had unemployment rates of 6.1 and 6.7 percent, respectively; Montgomery County, north and west of the city, had an unemployment rate of 5.0 percent, and the rate in Arlington County in Virginia (also west of D.C.) had a rate of just 3.6 percent. As the report's authors put it, "Washington is a region divided."[55]

That line separates many of the city's haves, who live in the city's western and northwestern areas, from its have-nots, who tend to live in the east and southeast. (In February 2013, the city's four eastern-most wards had the highest rates of unemployment.)[56] The latter often lack money or resources to move to the suburbs and the skills needed to obtain the area's many high-paying jobs.[57] Combine this with a possible influx of homeless people from neighboring states who seek city services,[58] and the result is nearly unparalleled economic segregation and inequality. In 2010, the richest 5 percent of D.C.'s households made nearly $475,000 per year, far more than any other large city, while the poorest 20 percent made just over $9,000 per year—one *fiftieth* as much. Only two American cities, Boston and Atlanta, had higher levels of income inequality. The January 2011 unemployment rate in Washington's poorest ward, Ward 8, which lies south of the Anacostia River, was a whopping 25 percent—higher than any metropolitan region of comparable working population, and rivaling the unemployment rates of economically crippled countries such as Spain and Greece in the early 2010s.[59]

In recent years, gentrification has been altering some of those eastern and southern neighborhoods, bringing in more people with education and wealth. But their arrival also threatens to squeeze out poorer residents who cannot afford higher rents and costs of living. Often, this is closely associated with racial divisions, as wealthier nonblack residents move into areas that have traditionally been African American. Black Washingtonians have long been on the losing end of city development, a result of the capital's failure to balance racial justice with economic growth. Since the 1950s, many have been forced by rising housing prices or even the wholesale destruction of existing housing stock to move out of neighborhoods like Dupont Circle, Capitol Hill, and areas in the southwest. In front of some homes in the historically black Anacostia neighborhood of Washington, located in one of the poorest wards in the city, "for sale" signs have been defaced by graffiti reading "no whites." Local political consultant Marshall Brown complained bitterly that "the new people believe more in their dogs than they do in people."[60]

On the other hand, it is also true that the city's leaders—including its African American mayors—have regularly supported the renewal of city neighborhoods. In fact, the appearance of new, nonblack residents in neighborhoods like U Street starting in the 1990s began well after large numbers of African Americans had left Washington for the suburbs.[61] Gentrification is also often color-blind. High school administrator and author Courtney Davis is one of many black professionals who have deliberately chosen to live in Anacostia. "I'm fighting for this neighborhood," she told one reporter. "It still has some work to do. But I'm not here to make a quick buck and run off." Joked another new Anacostia resident originally from California, "There are different types of people here, but that doesn't water down the chocolate."[62]

There is also reason to question whether the costs of local government policies, designed to encourage more economic activity, outweigh the benefits. Tax abatements may bring a new firm to the city, but it also results in lost tax revenue and does not always lead to productive or efficient development. Studies have shown that building a sports stadium rarely results in net economic benefits, yet Washington agreed to foot the entire bill for a new $600 million field for their major league baseball team. (In fact, of fifty-five sports stadiums built between 1997 and 2012, Washington was one of just seven that paid the full costs of construction.[63]) Finally, development in the city and the region has strained existing resources and infrastructure. The backed-up toilets of LeDroit Park are but one example. New nightclubs in NoMa have required more police patrols to respond to robberies and assaults.[64] Another example is traffic: despite the presence of Metro, thousands of people still drive to and from work every day, clogging the area's already-full roadways. One study found that the region has the single-worst traffic in the entire country, with commuters spending up to seventy-four hours per year on average stuck in traffic (versus thirty-four hours on average nationwide). Despite the construction of a new Metro line through Tysons Corner to Dulles Airport and the widespread use of informal carpooling (known as "slugging"), traffic problems continue to hinder productivity and make the area less attractive to would-be employers.[65]

Conclusion

As home to the nation's government, Washington can boast of a largely secure source of jobs and commerce that helps protect the city from national recessions. More than offering economic security, however, the federal government has encouraged major new developments in D.C., especially in the past decade. Today's Washington is, in many ways, a city of construction cranes. Other American cities have revived as well in recent years, a trend that suggests not only that urban life is more popular in general but that Washington is more economically diverse and similar to the typical urban center in America than people realize. Of course, growth is never permanent, and D.C.'s history is full of booms and busts. But even if the capital's current growth slows down, it seems unlikely it will return to the difficult era of the 1960s and 1970s anytime soon.

One of the enduring questions for the city's economy and the improvement of its neighborhoods is whether wealthy Washington is what two scholars have argued is the difference between a "boutique" city and a "Potemkin" city—that is, whether its rich residents are part of a truly thriving urban area, or if Washington's revival conceals serious economic and social problems.[66] There is certainly evidence for both. Newly revitalized neighborhoods have brought greater traffic and livelihood to places such as H Street, N.E., and Columbia Heights, and brand-new neighborhoods like NoMa have sprung up too. But much of the area's wealth is possessed by people who do not actually live in the city. Swaths of the district remain neglected and suffer from persistent crime and poverty. And as city property has become more valuable, areas that house the poor and needy—who often form communities of their own—have become too expensive for those citizens to live and work in. It is an open question whether Washington will overcome the tremendous challenge of translating its economic success into improved conditions for all of its citizens, not only the rich and well-educated.

CHAPTER 10

Neighborhoods and Suburban Communities of Washington

As the clock strikes noon on a sunny day in April, a steady stream of individuals moves in and out of the Eastern Market Metro station. Serenaded at the top of the escalator by a musician playing a bongo drum accompanied by a synthesized instrumental track, the eclectic travelers range from business suit-wearing men and women to tourists juggling cameras, strollers, and shopping bags. The surrounding streets are lined with narrow nineteenth-century row houses, many of which accommodate businesses on the street level with residences above. The market, for which the Metro stop is named, was built in 1872 and restored after a ruinous fire in 2007. It anchors the neighborhood and serves as both a community gathering place and a tourist destination. Looming large on the horizon is the U.S. Capitol building, leaving no doubt about why the neighborhood is known as Capitol Hill.

Only a few miles away, the northeast neighborhood of Brookland has a very different atmosphere. The main drag of Twelfth Street is flanked by small businesses on both sides. Detached single-family homes with both front- and backyards are intermixed with several apartment buildings, and the Brookland-CUA Metro station makes downtown easily accessible. Not exactly a tourist destination, the neighborhood has more of a small-town feel. Influenced heavily by nearby Catholic University of America and Howard University and known as "Little Rome" because of the high concentration of religious houses and churches, the area is racially diverse.[1] Church bells echo through the air to mark the middle of the day. The sounds of construction can also be heard, and several large buildings in varying states of completion mark the beginning of a new phase in the neighborhood.

These two areas, separated by only a few miles, provide a small glimpse of the neighborhood diversity found within Washington, D.C. The District of

Columbia boasts dozens of neighborhoods that vary widely in their culture, background, and identity. Similarly, outside of the city's boundaries are scores more suburban neighborhoods and towns in Virginia and Maryland that are also distinctive, each in their own way.[2]

Washington, D.C., is no different from other large American cities with neighborhoods and suburban communities. Like in many other urban places, the creation and identity of Washington's individual neighborhoods was driven in particular by transportation innovations and the specific locations of powerful institutions such as universities and government agencies. At the same time, however, Washington, D.C., has several distinct features that have affected neighborhood growth and development in the region. L'Enfant's plan for Washington influenced how and where neighborhoods would emerge in relation to the central city. The unique history of African American Washingtonians influenced the growth and development of specific neighborhoods. And the city's lack of local autonomy for much of its history encouraged the rise of strong community organizations and effective civic activism within neighborhoods and across the city that enhanced neighborhood identity.

Developing into a Residential City

There is a common misperception that the population of Washington, D.C., and the greater metropolitan area is comprised primarily of "rootless residents" who are only here because they are somehow connected to the government.[3] The comments of prominent Washingtonians do little to dispel this impression. Richard Nixon, for instance, once described Washington as a "city without identity," claiming that "everybody comes from someplace else."[4] If this were true, it would be hard to imagine how the area could develop durable neighborhoods and communities. But as we noted in chapter 5, the city is no more transitory than other American cities; and though it does have some temporary residents, many of whom having moved to the area specifically because they work for the federal government, the Washington, D.C., region also has a sizable native population. This combination of people influences how the city and its neighborhoods have developed and changed over time.

According to the 2010 U.S. Census, the population of the greater D.C. metropolitan region ranks seventh in the nation. This status as one of the largest urban centers in the country is a recent development. Washington's population grew relatively slowly during its early years. Carl Abbott, for instance, describes the capital as a "city built in the later decades of the twentieth century." Though Washington is over two hundred years old, he points out that "metropolitan Washington has added nearly 80 percent of its population and nearly 90 percent of its developed area since 1940." But while there has only recently been a "dramatic shift in the image of the district as a place to live as well as work," the city of Washington has in fact functioned in a residential capacity since its inception and has experienced periodic times of considerable growth coinciding with significant national events.[5]

The City's Growth and the Development of Its Neighborhoods

When George Washington selected the site for the new capital city, the population in the area that would become the District of Columbia was small. Mostly farmland, the region featured several large estates and the minor established settlements of Georgetown and Alexandria. Only after the district was established and the need for people to construct government buildings and, later, work in those buildings increased did the area's population begin to grow significantly.

As we have noted in previous chapters, L'Enfant's vision for Washington was a *living* city where people would settle and thrive, not merely work or stay temporarily. To fulfill this vision, the French architect included in his proposal the construction of a church, markets, a theater, and a bank—in other words, the services and amenities that any city's residents would need and desire. L'Enfant hoped that development would occur around the city's many squares, leading to the "emergence of separate communities in great open spaces." All would eventually connect, in L'Enfant's vision, but they could also conceivably become the focal points for the creation of city neighborhoods.[6]

Some of the earliest housing developments in the city were boarding houses, local taverns, and hotels that provided short-term housing options for government workers and elected officials who were only part-year residents or who considered their stay in Washington, D.C., temporary. The area around the Capitol quickly became a central location for such establishments. Like the residents who stayed in them, these businesses experienced a lot of change, often shifting locations from year to year or transferring ownership.[7] As the federal government gradually grew in responsibility and acceptance, however, so too did the need for permanent employees and support personnel in the city. These workers needed places to live. Small houses and neighborhood clusters began to emerge by the early 1800s. Each neighborhood reflected the nature of the types of businesses and services provided by the people settling there.

Not surprisingly, one of the first residential areas to develop within the new Washington City was the Capitol Hill neighborhood.[8] This area, which (as the name suggests) immediately abuts the U.S. Capitol building, included several distinct sections, and housing was built that ranged from ornate mansions to tenement houses. In his *Personal Reminiscences*, Dr. Samuel C. Busey paints a vivid portrait of this division, describing the residents of one section of Capitol Hill in the mid-nineteenth century as a collection of "quiet, church-going people of high social standing" including government officials, officers in the armed forces, attorneys, and scientists. Just a few blocks away, in what he deemed an "uninhabitable" tenement district located on Second Street between A Street and Maryland Avenue, Busey characterized the residents as "the most disorderly, drunken, and debased group of men, women and children, white and colored, that ever afflicted any section of this city." Residing in "filthy and lousy lodgings," these residents participated in "daily and nightly carousals" that were "disgusting, obscene, and unsafe."[9] Busey's colorful descriptions illustrate the diversity found within individual neighborhoods even in the district's early decades. The Capitol Hill neighborhood was one of the first areas of the city to develop into numerous "regions" distinguished by a broad range of socioeconomic features, and this pattern would be repeated throughout Washington.

As the city's infrastructure increased and businesses were built up to support the daily functioning of the government, expansion in the city center encouraged residential growth around Washington. The Seventh Street Corridor is an example of these developments. Functioning primarily as a residential area at the beginning of the 1800s, the neighborhood had changed significantly by the middle of the century and included a collection of federal and municipal buildings, residences, and small businesses.

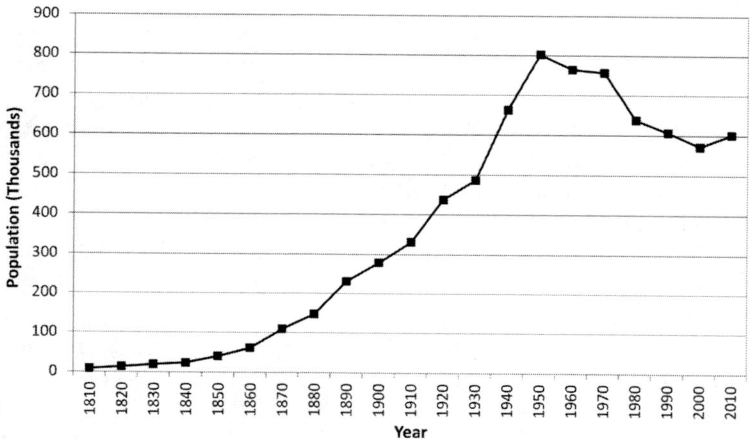

Figure 10.1 Population of Washington D.C., 1810–2010

Source: Gibson and Jung 2005 (1810–1990); U.S. Census, Population Division (2000–2010)

Between 1810 and 1850, the city's population swelled nearly fourfold, from 8,000 to 40,000 citizens (see figure 10.1). Neighborhoods began to develop as well—some along the lines envisioned by L'Enfant, others not. But the rise of new neighborhoods differed in at least one way from the same phenomenon in other American cities. In places like Boston, Chicago, and New York City, many individual neighborhoods were originally separate cities and towns that were incorporated into the larger metropolis; their initial independence encouraged them to maintain their unique community identities.[10] This was not quite what happened in Washington, which was originally a small city surrounded by the large rural "county" of the District of Columbia. Though Georgetown, which was made part of the city of Washington in 1871, was originally its own self-governing entity, most of the city's growth either led to the creation of new neighborhoods out of rural areas (such as Columbia Heights) or simply spread around and merged with existing but unincorporated communities within the larger district (such as Tenleytown, in the upper northwest quadrant). As the urban core expanded, these areas had to find a way to balance an identity of being part of Washington, D.C., with their own individual community character.

City growth continued steadily through the rest of the nineteenth century, spurred by the influx of free blacks during the Civil War. (The city's population grew by 79 percent between 1850 and 1860—its highest rate of growth of any decade in its history.) By the 1920s, the city finally reached the borders of the District of Columbia, but growth did not halt: instead, new suburban communities emerged in places including Silver Spring, Maryland, and Arlington, Virginia, which grew enormously in population and economic development. The New Deal and World War II brought tremendous expansion to the national government and, with it, an influx of more residents; the city's population increased by nearly two thirds, from 487,000 in 1930 to 802,000 in 1950, but many more found places to live beyond D.C.'s borders. Even the federal government tried its hand at creating a new suburban community with the founding of Greenbelt, Maryland, in 1937 to provide affordable housing in the D.C. area.[11]

Some of these new residential settlements outside the official city limits, like Greenbelt, have been more successful than others at developing community identity. But what they all had in common is that D.C., because of its fixed boundaries, could never be expanded to include them. So whereas other cities, at least through the early twentieth century, have often grown by incorporating their suburbs into their borders, Washington's outer communities are assured that they will remain parts of their respective states.

The Role of Transportation Systems

No city can grow when traveling distances are too great. Washington, D.C.'s attractiveness to settlers and periodic need for more government and service workers has been the obvious prerequisite for its growth. But like other urban areas, transportation innovations have been critical to the city's expansion and the creation of new residential and suburban communities.

One of the earliest forms of public transportation in Washington was the horse-drawn streetcar. Beginning regular service for the public in 1862, it operated on tracks that generally followed main roads. As urban historian Kenneth Jackson points out, these tracks often radiated out from the center of the city in a linear fashion. Offering quick, efficient transportation along set routes, they "were developed toward the emerging wealthy neighborhoods on the periphery" and "enhanced property attractiveness."[12] This innovation encouraged new neighborhoods to develop. Known as "streetcar suburbs," these residential areas grew up along the streetcar tracks.

Mount Pleasant, one of the earliest suburbs of Washington City, developed in the northwest quadrant of the District of Columbia as a result of the horse-drawn streetcar. Located at the end of one of the city's main streetcar lines, this turnaround spot provided a prime location for residential and business development. A commercial corridor flanked by large, single-family homes, row houses, and apartment buildings was rapidly built in the neighborhood between 1900 and 1930.[13]

Horse-drawn vehicles were eventually replaced with electric streetcars and commuter railroads, further encouraging the outward movement of population. Offering speeds up to four times faster than horse-drawn streetcars, these new forms of transportation provided expanded options for travelers in the city and led to the development of residential areas even further from the city core. Operating along set routes, streetcars and commuter railroads provided individuals with regular transportation to and from the central business core of the city both for work and leisure activities, connecting the outlying residential communities beyond reasonable walking distance to downtown. The routes were generally designed to connect existing villages and neighborhoods to the city center, but, once the lines were constructed, new neighborhoods emerged along the tracks, while existing ones tended to grow in size. Brookland, for example, expanded from a sleepy village to a streetcar suburb when a commuter railroad, and later a streetcar line, was built in the area. Chevy Chase, by contrast, was planned with the intention of constructing a new streetcar line to provide access between the neighborhood and downtown Washington.[14]

The widespread accessibility of the automobile by the 1920s allowed development to continue in an even more dramatic fashion, as growth was no longer limited to the space along established tracks. New roads were built to make existing residential areas more accessible, and additional neighborhoods emerged in areas that were previously undeveloped.[15] As a result, new "automobile suburbs" began to appear further from the center of the city. Located primarily outside of the boundaries of the District of Columbia in northern Virginia and southern Maryland, these neighborhoods generally had lower population densities and larger lot sizes than urban neighborhoods or the streetcar and trolley neighborhoods. As businesses, industry, government offices, shopping malls, and entertainment venues opened and thrived in these suburbs located along the urban fringe, they became places of work as well as places of residence. This new category, labeled "edge cites" by Joel Garreau, combined elements of both urban and suburban communities.[16] Tysons Corner, Virginia, and Silver Spring, Maryland, are examples of automobile suburbs that have become such "edge cities" (see textbox 1).

TEXTBOX 1: Tysons Corner, Virginia

Tysons Corner is a quintessential example of how new transportation technology has dramatically altered—and continues to alter—the shape and identity of communities in the Washington metropolitan area.[17]

Tysons Corner is an approximately four-mile-square section of land in Fairfax County, Virginia, ten miles west of Washington, D.C. The first transportation innovations in the area were railroads, canals, and roads built during the Civil War, but these did little to bring development. It was not until the 1950s that the region began to transition from a primarily rural area to a business district—the direct result of two developments: new road construction and increased defense spending during the Cold War.

With the construction of the Beltway (I-495) and the Dulles Access Road, which allowed for vehicular transportation to the then-new Dulles International Airport, Tysons Corner—which rested just southwest of where the two roads intersect—grew significantly. Tysons Corner also benefitted from the construction of I-66, which provided a direct route from D.C. to areas in Virginia due west of the city, including Tysons Corner. Zoning laws in the area encouraged real estate development, residential housing, and suburban business development that coincided with the growth of companies doing business with the Department of Defense. The department itself had relocated to its new Pentagon facility in northern Virginia in 1943.

(*continued*)

In recent decades Tysons Corner became a popular shopping destination as well. Its Tysons Corner Center, completed in 1968, is an enormous mall that boasts over two million square feet of space. It has also proven popular as a center of telecommunication and Internet firms, and the resulting traffic congestion has become a major problem in the primarily automobile-oriented Tysons Corner.

A new form of transportation—the Washington Metro system—promises yet more changes to the area. The Metro is being extended through Tysons Corner as part of the new Silver Line. The Fairfax County Board of Supervisors, meanwhile, approved plans to control development near Metro stations, allow taller buildings around these stations, and develop more pedestrian-friendly streets and mixed-use buildings with street-level businesses. This "town-center development," which can also be seen in Bethesda, Leesburg, Rockville, and other suburban communities around Washington, D.C., is designed to replicate the culture and convenience of city life without some of its drawbacks, such as traffic congestion and crime. It has even instigated a rebranding of the neighborhood: local leaders have begun promoting a new name for the area—leaving off the "Corner" and calling it simply Tysons—perhaps to shed its reputation as a shopping mecca and make it sound less provincial.[18]

This new development does not come without challenges. Since Tysons Corner is a well-developed area designed for high volumes of vehicular traffic, "retrofitting Tysons into an urban street grid is a challenge on a scale that urban planners and academics say they have never seen."[19] But if successful, the project promises to inaugurate a bold new chapter in the history of Tysons Corner.

The rapid rise in automobile use caused growing concern about traffic congestion and parking in the Washington, D.C., metropolitan area. In most cities around the country, the initial solution was to build yet more roads. Robert Moses, city engineer and planner for the city of New York, was one of the first successful advocates of this approach, resulting in new freeways and expressways that encircled, and sometimes cut directly through, New York neighborhoods.[20] When the federal government enacted the Federal Aid Highway Act in 1956, which paid 90 percent of the cost of new freeways, cities and states had a huge financial incentive to build additional roadways. Construction for the Capital Beltway, a freeway that encircled Washington, began in 1957 and was completed in 1964. As traffic in D.C. continued to grow, still more freeways were envisioned. In 1959, an extensive freeway system known as the Mass Transportation Plan was proposed for the city that would have required the demolition of numerous residential areas. But by then, resistance to such freeways was emerging around the country, including in the district. Concerned citizens rose up to protect their neighborhoods, and much of the plan was never implemented. In its place a rapid-transit system was suggested, which found wider support from citizens around the region.[21]

Offering a commuting alternative to the automobile, but without the same devastating effect on existing neighborhoods, the rapid-rail Metro system began operation in 1976. The complex and contested story of its planning and implementation includes several major themes connected to neighborhood identity. The desire to preserve neighborhoods slated for demolition for freeways led to the organization and empowerment of neighborhood groups; regional cooperation across political jurisdictions emerged in order to build, maintain, and manage the Metro system; and some declining neighborhoods around transit stations were renewed.[22]

This last theme has been particularly important for D.C.-area neighborhoods. The location of Metro stations provided an opportunity for some communities both inside and outside of the city's borders to reestablish a commercial center or otherwise encourage local development around the station. This kind of "transit-oriented development" has often required cities to issue mixed-use zoning regulations that encourage simultaneous business and residential growth. Street-level businesses with residences on upper floors help create a pedestrian-friendly feel to neighborhoods and eliminate the need for local residents to have cars.[23] Some neighborhoods within the District of Columbia, such as Columbia Heights, have anchored their own revitalization or gentrification plans around the location of Metro stations. Though not a sufficient spur for immediate development—some stations, like NoMa-Gallaudet and Wheaton, Maryland, have only recently become epicenters of new development, decades after their construction—the Metro has, similar to transportation of other kinds in the past, clearly been a major factor in determining what gets built where (see textbox 2).

TEXTBOX 2: Arlington, Virginia

Located in northern Virginia along the Potomac River, Arlington County was connected to Washington, D.C., by the construction of an electric streetcar line between Ballston and the central city in 1896. This connection led to growth and development in the area. Hoover Field was built on the present site of the Pentagon in the 1930s, and the area was popular for residential development.

During the Cold War, increased defense funding impacted the area enormously. The county housed numerous defense and military offices, and thousands of government-related jobs were created in the area. The housing market was unable to keep up with increased demand, forcing many individuals and families to live outside of the county. In the 1970s, Arlington County had a 10 percent population loss.

The proposed expansion of the Metro transit system into the greater metropolitan area provided Arlington County an opportunity to redefine itself and promote new growth. The new Metro stations were soon surrounded by new multi-use buildings that could house integrated businesses, stores, and residences. Unlike some of the Metro stations further west, Arlington County did not propose large parking lots by their transit stations. Instead, the areas near the stations were designed to support primarily pedestrian access in the hopes that people would work, shop, and engage in leisure activities near their residences while having easy access to downtown Washington, D.C. The local government utilized special permit processes to promote the desired high-density, mixed-use development within half a mile of Metro stations. They required that residential development occur in conjunction with office and business development and limited the construction of parking lots and garages.

The plan was quickly deemed a success. The population in Arlington increased 24 percent in the 1980s and 1990s. Condominium ownership increased and housing values skyrocketed. By 2000, the relative housing value in Arlington was 147 percent of the national average, the highest in the D.C. metropolitan area. By concentrating new growth near public transit, Arlington was able to invigorate redevelopment and redefine its community by achieving a balance between walking, mass transit, motor vehicles, and a combination of public and private activities.[24]

The African American Experience

Neighborhoods in Washington, like in other cities, frequently developed and have been characterized along racial and ethnic lines. For example, the working-class community along the waterfront in southwest D.C. welcomed large numbers of Italian and German immigrants in the first half of the nineteenth century, and sections of the Adams Morgan neighborhood were first primarily occupied by middle-class Jewish merchants in the 1800s, then became a hub of Latino culture and business in the twentieth century. While many of these ethnic neighborhoods developed as a result of particular preferences, some groups, particularly African Americans, were forced to live in certain areas because of racism and segregation.

Washington, D.C., home to a vibrant African American population, remained racially segregated well into the twentieth century, as was the case in so many American cities. Though this segregation denied blacks the right to settle where they wished, it did strengthen the identity of those neighborhoods where they were allowed to live. This encouraged the creation of a complex economic, social, and political network (the "secret city" mentioned in chapter 8) among African Americans in Washington, D.C.[25]

The U Street neighborhood in northwest D.C. provides one particularly impressive example of this distinct community identity. Growth in the neighborhood followed public streetcar lines after the Civil War. Initially attracting residents of diverse economic backgrounds, this area was widely recognized as a predominantly African American community by the beginning of the twentieth century. Racial segregation and discriminatory housing practices throughout the city led to development of this "city within a city." Serving as both a symbolic and physical gathering place for many African American residents of the city, it was a central location for businesses built by and for African Americans. Festivals, concerts, and sporting events further solidified the sense of community within the neighborhood.

When the Supreme Court deemed neighborhood segregation unconstitutional, racial barriers came down. Many residents left the area for newly accessible neighborhoods that were less crowded. The identity and reputation of African American neighborhoods like U Street also began to change, especially with rising racial tensions during the civil rights era and the race riots following Martin Luther King, Jr.'s assassination in 1968. The once vibrant, thriving neighborhood was labeled "unsafe," and the local economy suffered as the area became known for drug activity and many businesses left.

More recently, however, the U Street area has experienced new growth. Revitalization began with two key events: the decision to build the Reeves Municipal Building in 1986 and the construction of a Metro station in 1991. Since then, historic buildings have been restored and museums and memorials throughout the area have been built to commemorate people and places within the neighborhood that have contributed to Washington's African American identity within the city.[26]

Like the U Street area, other D.C. neighborhoods have historically had a strong African American identity, including LeDroit Park, Brightwood, the Shaw District, and Anacostia. In some of these places, including U Street, new development is seen as potentially undermining African American neighborhood identity (such as the attempt to rename an intersection on U Street "Little Ethiopia," mentioned in chapter 8). But attention has also been given to the black history in other neighborhoods around the city where it has often been overlooked, such as Georgetown. A documentary and a book, both entitled *Black Georgetown Remembered*, have brought much of this history to the forefront, celebrating the important role of black businesses, churches, and the arts in Georgetown.[27]

In short, African Americans have made significant contributions to numerous individual neighborhoods and communities throughout the city. Prominent black artists, intellectuals, and leaders in the struggle for civil rights have proudly called Washington, D.C., their home. As a result, the African American community in Washington, D.C., is an integral part of many of the capital's neighborhoods.

Community Action Groups

A large and prominent black population is not the only feature of Washington that distinguishes its neighborhoods' identities and growth from those of other cities. The unique political status of the capital has helped promote unusually strong community action groups and encouraged individuals and associations to fight for their neighborhoods.

Because there were no locally elected officials in Washington, D.C., from the mid-1870s until the mid-1960s, civic and citizen associations emerged as an alternative venue for city residents working together to advocate for services that would improve their neighborhoods. The leaders of these voluntary community organizations often filled the role that local elected officials had in other cities.[28] Neighborhood organizations worked to improve their communities in a variety of ways. They advocated for emergency services, such as fire protection and ambulance routes; called for transportation improvements, such as expanded streetcar lines, increased miles of railroad tracks, more bridges, and the paving of roads; and campaigned for funding for community services, such as libraries and schools.

Neighborhood groups also worked to stop city plans that they felt would harm their communities. One example of this kind of collective campaigning can be seen in the anti-freeway movement of the 1960s and 1970s. Initiated by citizens of the Brookland neighborhood of D.C. and the town of Takoma Park, Maryland, this movement gained support from neighborhoods around the city as residents sought to protect their respective communities from demolition to make room for multilane highways. Though not successful in all areas, it garnered considerable public attention and was even able to stop new freeways in some areas (see textbox 3).

TEXTBOX 3: The North Central Freeway Project

After WWII, as buses began to replace streetcars and automobiles became a part of everyday life in the greater metropolitan area, new threats to communities arose. The North Central Freeway (NCF) Project, which was part of a large network of freeways designed to address the needs of commuters coming into the city from the expanding suburbs, provided the impetus for a grassroots civic movement against highway development and for the protection of neighborhoods and residential areas that were slated to be demolished. Starting in Takoma Park and Brookland, the movement would eventually gain support throughout the 1960s and 1970s.

Specifically intended to connect Silver Spring, Maryland, to downtown Washington, the planned route of the NCF would go directly through both Takoma Park and Brookland and require the demolition of several hundred homes. Many residents in both neighborhoods were strongly opposed to the freeway project and worked diligently to prevent its construction. This fight, which began in 1961, included efforts by neighborhood associations in both neighborhoods. The associations began working together to present a unified opposition, and they gained support from civic groups in other parts of the city. Inter-neighborhood coalitions were formed, including both the citywide Emergency Committee on the Transportation Crisis and the Citizens' Committee on the Freeway Crisis. In addition to rejecting the plans for the freeway, many community members also supported alternative transportation solutions, including a rapid-rail transit system.

The NCF project was effectively stopped, largely through the efforts of neighborhood committees and citywide civic activism. In the early 1970s, the city announced that funds originally designated for the freeway would be used to develop a rapid-rail instead.[29] However, some neighborhoods were not as fortunate as Takoma Park and Brookland. Parts of Georgetown, Foggy Bottom, Kenilworth, and the southwest portion of the city were demolished to make space for highways. Houses were torn down, local streets were redirected, and local businesses were hurt.[30] Still, the curtailment of the NCF illustrated the potential power of neighborhood activism and organization in Washington.

Neighborhood associations have remained important even after the threat of destructive new freeways gradually disappeared. They have been successful, for instance, in protecting historic buildings and preserving and celebrating the history and cultural heritage of their communities.[31] When D.C. began exercising home rule in 1974, the lasting impact of neighborhood organizations became even clearer. That year a district referendum (mandated by the federal law that granted home rule) established a unique element of the local government known as Advisory Neighborhood Commissions (ANCs). These 200-plus commissions are made up of locally elected officials, and each one represents an individual neighborhood within the District of Columbia. In addition to having a voice in the district's budget, ANC representatives consider issues that directly impact their own neighborhoods, such as public transportation, street improvements, police protection, trash collection, and local zoning questions. They are allowed to present their positions to district government agencies, the city council, the mayor, and even testify before federal agencies when appropriate. The ANCs demonstrate the lasting legacy of powerful community organizations within the district and contribute to the local identity, culture, and heritage of city neighborhoods.[32]

In addition to the ANCs, other types of local neighborhood organizations preserve the heritage of individual neighborhoods and foster strong community ties. Interview collections, such as the Voices of 14th Street series and the Hillcrest Community Civic Organization oral history project, capture the stories of communities through the voices of residents and preserve community history.[33] Associations and civic organizations also exist to support local commercial interests. With activities as diverse as maintaining community blogs and sponsoring farmers' markets and local festivals, these associations allow their neighborhoods to maintain, enhance, and celebrate their own unique identities.

Conclusion

Each neighborhood of Washington, D.C., is a vibrant part of the living city. It is where people live, work, and develop a communal identity. Each has developed individually in response to the growth of the federal government, to transportation innovations, and to housing needs (and, in the case of African Americans, restrictions). Neighborhood organizations and community action groups have been, and continue to be, powerful forces in local politics, helping reinforce local identity. And identity has been just as important for many communities outside of the city of Washington, which also have changed in response to government policies and new forms of transportation.

While residents in the greater metropolitan region of Washington, D.C., do often identify with their local community, there are also important ties between them. As noted in the previous chapter, the area is economically interconnected; the jobs of thousands of Marylanders and Virginians are affected by the budgetary decisions of the federal government. People in the region travel on the same Metro lines, cheer for the same "home" sports teams (though some Marylanders prefer to root for Baltimore teams), and often identify themselves as "Washingtonians." In addition, complex cross-jurisdictional relationships have been forged in the greater metropolitan region. Organizations such as the Metropolitan Washington Council of Governments, the Greater Washington Board of Trade, and the Washington Metropolitan Area Transit Authority (which oversees the Metro) have helped to create and reinforce a single regional identity.

It has not always been easy for D.C. neighborhoods and suburban communities to preserve individual identity amid these homogenizing forces. New developments in older neighborhoods can also threaten their architectural and social uniqueness. But many still find ways to maintain and celebrate their identity, even if only superficially. Chinatown's chain stores are required to have signage in Chinese script, for instance. And in the midst of a large development project in Brookland that includes new apartment buildings, artist studios, businesses, and community spaces, one of the new buildings has been adorned with the name of the neighborhood in huge block lettering. This community "branding," easily visible from the Metro platform, solidifies the community's identity and marks continuity in the midst of change. These examples are evidence that Washington, D.C., is as much a collection of places as it is a single urban metropolis.

Notes

Introduction

1. See, for example, Kip Lornell and Charles C. Stephenson, *The Beat! Go-Go from Washington, D.C.* (Jackson, MS: University Press of Mississippi, 2009), 2.
2. Howard Gillette, Jr., for instance, argues that national politics (namely, national urban policy) affected local urban planning, with both advantages and disadvantages for the city, including its prominent black population. Howard Gillette, Jr., *Between Justice and Beauty: Race, Planning, and the Failure of Urban Policy in Washington, D.C.* (Baltimore: Johns Hopkins University Press, 1995), x–xi. We argue, however, that these three features of the city extend beyond urban policy to other spheres as well.
3. Alan Lessoff, *The Nation and Its City: Politics, "Corruption," and Progress in Washington, D.C., 1861–1902* (Baltimore, MD: Johns Hopkins University Press, 1994), 1; James H. S. McGregor, *Washington: From the Ground Up* (Cambridge, MA: Harvard University Press, 2007), 297.
4. Carl Abbott perhaps puts it best when he writes that Washington has had both a vertical dimension (its national and international identity) and a horizontal dimension (its regional identity). Carl Abbott, *Political Terrain: Washington, D.C., from Tidewater Town to Global Metropolis* (Chapel Hill: The University of North Carolina Press, 1999), 6.
5. Mary Meade Coates, quoted in Jill Connors, ed., *Growing Up in Washington, D.C.: An Oral History* (Charleston, SC: Arcadia Publishing, 2001), 34.
6. Margaret Farrar, *Building the Body Politic: Power and Urban Space in Washington, D.C.* (Urbana: University of Illinois Press, 2008), 13.
7. Abbott 1999, 65–66, quotes pp. 7–8.
8. Our thanks to Tim Meagher for his conceptualization of Washington as a set of "cities."
9. Abbott (1999), for instance, writes extensively about how the city's recreational history has stemmed from its regional identity.
10. For a recent review of these schools of urban theory, see Dennis R. Judd, "Theorizing the City," in *The City, Revisited: Urban Theory from Chicago, Los Angeles, and New York*, eds. Dennis R. Judd and Dick Simpson (Minneapolis: University of Minnesota Press, 2011).

11. Kevin Starr, *Golden Gate: The Life and Times of America's Greatest Bridge* (New York: Bloomsbury Press., 2010), 29.
12. In addition, the university has played an important role in charitable work in the city. See, for example, Jenell Williams Paris, "*Fides* Means Faith: A Catholic Neighborhood House in Lower Northwest Washington, D.C.," *Washington History* 11:2 (Fall/Winter 1999–2000).

Chapter 1

1. Technically, an avenue is a street "starting and ending at terminated vistas." Some avenues in Washington (most obviously Pennsylvania Avenue) fit this description, but most others (e.g., Hawaii) do not. Dhiru A. Thadani, *The Language of Towns and Cities: A Visual Dictionary (*New York: Rizzoli, 2010), 49.
2. Defined roughly as the area bounded on the northwest, north, and northeast by Florida Avenue; the east by Nineteenth Street NE/SE; the south by the Anacostia River; and the west by the Potomac River and Rock Creek.
3. Kostof, *The City Shaped*, 211.
4. These include Anglo Palladianism, Palladianism, Neo-Palladianism, Roman Revival, Greek Revival, Beaux-Arts Classicism, and Stripped Classicism. William Pierson offers an alternative categorization, with four distinct eras of neoclassicism in the pre–Civil War United States (Traditional, Idealistic, Rational, and National phases); see Pierson, *American Buildings*, 211.
5. Sources: Adam, *Classical Architecture*; Doreen Yarwood, *Encyclopedia of Architecture* (New York: Facts on File Publications, 1986), 128–142.
6. Mark Gelernter, *A History of American Architecture: Buildings in Their Cultural and Technological Context* (Hanover, NH: University Press of New England, 1999), 14; Ann Sutherland Harris, *Seventeenth Century Art & Architecture* (London: Laurence King Publishing, 2005), 244–46; Spiro Kostof, *The City Shaped: Urban Patterns and Meanings through History* (Boston: Bulfinch Press, 1993), 52, 214; John W. Reps, *The Making of Urban America: A History of City Planning in the United States* (Princeton, NJ: Princeton University Press, 1965), 108–114, 128–130, 157–174, 183–192, 205–207. Examples of planned American cities include Williamsburg, New Haven, Philadelphia, and Savannah; see Vincent Scully, *American Architecture and Urbanism*, revised edition (New York: Henry Holt and Company, 1988), 30–34.
7. Kenneth R. Bowling, *The Creation of Washington, D.C.: The Idea and Location of the American Capital* (Fairfax, VA: George Mason University Press, 1991). In a later essay, Bowling notes that the phrase *seat of government* was preferred over the word *capital* because the latter implied the centrality of the national government; it did not gain popularity until after the Civil War. Kenneth R. Bowling, "A Capital before a Capitol: Republican Visions," in *A Republic for the Ages: The United States Capitol and the Political Culture of the Early Republic*, ed. Donald R. Kennon (Charlottesville, VA: University Press of Virginia, 1999), 37.
8. Bowling, *The Creation of Washington, D.C.*, 29–34; J. L. Sibley Jennings, Jr., "Artistry as Design, L'Enfant's Extraordinary City," *The Quarterly Journal of the*

Library of Congress (1979): 230. For more on the prejudice of Founding Fathers and other Americans against large commercial cities as a home for government, see James M. Banner, Jr., "The Capital and the State: Washington, D.C., and the Nature of American Government.," in *A Republic for the Ages: The United States Capitol and the Political Culture of the Early Republic*, ed. Donald R. Kennon (Charlottesville, VA: University Press of Virginia, 1999); Bowling, "A Capital before a Capitol."

9. Carl Abbott, *Political Terrain: Washington, D.C. from Tidewater Town to Global Metropolis* (Chapel Hill: The University of North Carolina Press, 1999), 29, 34–35; Bowling, *The Creation of Washington, D.C.*, 4, 44, 123, 212–214; Elbert Peets, *On the Art of Designing Cities: Selected Essays of Elbert Peets*, ed. Paul D. Spreiregen (Cambridge, MA: The M.I.T. Press, 1968), 5. George Washington also had personal reasons for making this choice: his Mt. Vernon home was on the Potomac, he believed deeply in the river's economic potential, and he was president of a company seeking to connect the upper Potomac to the Ohio. See Scott W. Berg, *Grand Avenues: The Story of the French Visionary Who Designed Washington, D.C.* (New York: Vintage Books, 2007), 97–101; Bowling, *The Creation of Washington, D.C.*, 118–121. Abbott also notes that the border between North and South was not firm at the time. See Abbott, *Political Terrain*, 33.
10. Berg, *Grand Avenues*, 78–79, 84; Kostof, *The City Shaped*; Pamela Scott, "'This Vast Empire': The Iconography of the Mall, 1791–1848," in *The Mall in Washington, 1791–1991*, ed. Richard Longstreth (Washington, D.C.: National Gallery of Art, 1991), 39. L'Enfant had access to, and possibly consulted, plans of other cities in Europe and the United States and had previously lived in Versailles and Paris, which both had baroque elements of design. Comparisons have also been made between L'Enfant's plan and the post-sixteenth-century redevelopment of Rome, Paris (both as it looked and its proposed redevelopment), Karlsruhe, Versailles, a plan for rebuilding London after the 1666 fire, the design of a royal chateau in Marly (France), St. Petersburg, and Madrid. Earlier American cities, such as New Haven, Philadelphia, Annapolis, and Williamsburg, had certain features found in L'Enfant's proposal, as did early plans of the city drawn by Thomas Jefferson. Berg, *Grand Avenues*, 87–88, 106–112; Jennings, "Artistry as Design," 271–272; Kostof, *The City Shaped*, 211, 216; Iris Miller, *Washington in Maps: 1606–2000* (New York: Rizzoli International Publications, 2002), 42; Peets, *On the Art of Designing Cities*, 13, 20–24; Reps, *The Making of Urban America*, 22, 130, 172, 174; John W. Reps, *Monumental Washington: The Planning and Development of the Capital Center* (Princeton, NJ: Princeton University Press, 1967), 4–5, 8, 10, 15; and Scott, "This Vast Empire," 43–45.
11. Peets, *On the Art of Designing Cities*, 14, 43; Reps, *Monumental Washington*, 20.
12. Berg, *Grand Avenues*, 80, 103–104; Donald E. Jackson, "L'Enfant's Washington: An Architect's View," *Washington History* (1978): 403–405; Kostof, *The City Shaped*, 209–210; Peets, *On the Art of Designing Cities*, 32; Reps, *Monumental Washington*, 20, 22; Richard W. Stephenson, *A Plan Wholly New: Pierre Charles*

L'Enfant's Plan of the City of Washington (Washington, D.C.: Library of Congress, 1993), 28–29, 52.

13. Bowling, *The Creation of Washington, D.C.*, 6; Jennings, "Artistry as Design," 266; Lewis Mumford, *The City in History* (San Diego: Harcourt, 1961), 403; Reps, *Monumental Washington*, 19–20; and Stephenson, *A Plan Wholly New*, 54–58. Others also called for Washington to echo the great cities of the past; see, e.g., Bowling, *The Creation of Washington, D.C.*, 220.
14. Berg, *Grand Avenues*, 78–79; Jackson, "L'Enfant's Washington," 398, 409; Jennings, "Artistry as Design," 263–267; Reps, *Monumental Washington*, 18–20; and Stephenson, *A Plan Wholly New*, 59; see also excerpt from L'Enfant's description of the plan in Frederick Gutheim and Antoinette J. Lee, *Worthy of the Nation: Washington, D.C., from L'Enfant to the National Capital Planning Commission*, 2nd Edition (Baltimore: Johns Hopkins University Press, 2006), 23; see also chapter 9.
15. Vitruvius, *On Architecture*, trans. Richard Schofield (New York: Penguin Books, 2009), 19; see also Wayne Attoe, "Theory, Criticism, and History of Architecture," in *Introduction to Architecture*, eds. James C. Snyder and Anthony J. Catanese. (New York: McGraw-Hill, 1979), p. 32.
16. Mumford, *The City in History*, 404–406; Peets, *On the Art of Designing Cities*, 15, 38–39; Reps, *Monumental Washington*, 22. Thomas Jefferson himself worried about the odd-shaped blocks of the map, but they were still sold as potential development sites, even though L'Enfant suggested they be parks (Peets, *On the Art of Designing Cities*, 16). City Beautiful advocates, by contrast, believed that the odd blocks could "inspire a noteworthy public statement," as suggested by the triangle-based East Wing of the National Gallery (Kostof, *The City Shaped*, 235; Francis D. Lethbridge, "The Architecture of Washington, D.C.," in *The AIA Guide to the Architecture of Washington, D.C.*, ed. G. Martin Moeller, Jr. (Baltimore: Johns Hopkins University Press, 2006), 16). For additional criticisms of the plan, see Peets, *On the Art of Designing Cities*, 39–42; Reps, *Monumental Washington*, 22.
17. Jennings, "Artistry as Design," 268; Mumford, *The City in History*, 391–92, 401, 408. According to the architect Donald Jackson, L'Enfant actually wanted to achieve "a greater sense of immediacy and human scale within the urban core" via short streets, trees, and perhaps archways and arcades (Jackson, "L'Enfant's Washington," 406).
18. Berg, *Grand Avenues*.
19. Gelertner, *A History of American Architecture*, 120–123; William H. Pierson, Jr., *American Buildings and Their Architects, Volume One: The Colonial and Neoclassical Styles* (New York: Oxford University Press, 1970), 14, 287, 297, 332, 356; Damie Stillman, "From the Ancient Roman Republic to the New American One: Architecture for a New Nation," in *A Republic for the Ages: The United States Capitol and the Political Culture of the Early Republic*, ed. Donald R. Kennon. (Charlottesville, VA: University Press of Virginia, 1999). For more on Jefferson's architectural style and influence, see Charles E. Brownell, "Thomas Jefferson's

Architectural Models and the United States Capitol," in *A Republic for the Ages: The United States Capitol and the Political Culture of the Early Republic*, ed. Donald R. Kennon (Charlottesville, VA: University Press of Virginia, 1999); Buford Pickens, "Mr. Jefferson as Revolutionary Architect," *Journal of the Society of Architectural Historians* 34, no. 4 (1975).

20. Berg, *Grand Avenues*, 88; Gutheim and Lee, *Worthy of the Nation*, 41–42; Pierson, *American Buildings*, 398. See also Pamela Scott, *Temple of Liberty: Building the Capitol for a New Nation* (New York: Oxford University Press, 1995). Jefferson also employed a young Robert Mills at Monticello and later encouraged him to gain valuable training under Benjamin Latrobe (Pierson, *American Buildings*, 373–74). George Washington also was keenly interested in the design of the Capitol building. Kenneth R. Bowling, "'The Year 1800 Will Soon Be upon Us': George Washington and the Capitol," in *A Republic for the Ages: The United States Capitol and the Political Culture of the Early Republic*, ed. Donald R. Kennon (Charlottesville, VA: University Press of Virginia, 1999).
21. Kenneth Hafertepe, "An Inquiry into Thomas Jefferson's Ideas of Beauty," *Journal of the Society of Architectural Historians* 59, no. 2 (2000): 216–231; Pickens, "Mr. Jefferson," 26; Robert Tavernor, *Palladio and Palladianism* (London: Thames & Hudson Ltd., 1991), 192, 198. The architect of the Lincoln Memorial, Henry Bacon, also saw the memorial as having an instructional purpose, as did the landscape artists of the 1840s and 1850s who wanted to design the National Mall (Therese O'Malley 1991, "'A Public Museum of Trees': Mid-Nineteenth Century Plans for the Mall," in *The Mall in Washington, 1791–1991*, ed. Richard Longstreth (Washington, D.C.: National Gallery of Art, 1991), 63; Christopher Thomas, *The Lincoln Memorial and American Life* (Princeton, NJ: Princeton University Press, 2002), 65).
22. Some have argued that the plan contains more covert, often Masonic, symbols, but there is no evidence that L'Enfant, Washington, or anyone else had such intent. For a critical review of writings in this category, see Don Alexander Hawkins, "Masonic Symbols in the L'Enfant Plan: An Examination of Recent Publications," *Washington History* 21 (2009): 100–105. Other meanings have also been found in Washington's architecture; see, e.g., Scott, "This Vast Empire."
23. Mumford, *The City in History*, 406; Scott, "This Vast Empire," 37; James Sterling Young, *The Washington Community, 1800–1828* (New York: Columbia University Press, 1966). Others have found symbolic meaning in the location of the city itself, the creation of state-oriented squares, the size of the Mall, and the naming of the city's streets (see, e.g., Bowling, *The Creation of Washington, D.C.*, 224; Richard Longstreth, "Introduction: Change and Continuity on the Mall," in *The Mall in Washington, 1791–1991*, ed. Richard Longstreth (Washington, D.C.: National Gallery of Art, 1991), 14; and Scott, "This Vast Empire," 39–40).
24. See, e.g., Gelertner, *A History of American Architecture*, 100, 114; Hanno-Walter Kruft, *A History of Architectural Theory* (New York: Princeton Architectural Press, 1994), 345; Tavernor 1991, 188. Democratic thinkers in eighteenth-century

England were more explicit in seeing democratic meaning in neoclassical architecture (Gelertner, *A History of American Architecture*, 100).

25. Gelertner, *A History of American Architecture*, 116; Spiro Kostof, *The City Assembled: The Elements of Urban Form through History* (London: Thames & Hudson, 1992), 195; Jeffrey F. Meyer, *Myths in Stone: Religious Dimensions of Washington, D.C.* (Berkeley, CA: University of California Press, 2001), chapter 2; Mumford, *The City in History*, chapter 13; Vincent Scully, *American Architecture and Urbanism*, revised edition (New York: Henry Holt and Company, 1988), 72; Vincent Scully, *Architecture: The Natural and the Manmade* (New York: St. Martin's Press, 1991), 84, 117; Tavernor, *Palladio and Palladianism*, 93. The association of neoclassicism with fascism has tainted such recent neoclassical constructions in Washington as the World War II Memorial (Moeller, *AIA Guide to the Architecture of Washington, D.C.*, 88–89). The baroque city plan has been attractive to totalitarian and autocratic regimes for its "presumption of absolute power" (Kostof, *The City Shaped*, 217), but also because, practically speaking, military authority can more quickly reach multiple areas of a baroque city (Mumford, *The City in History*, 388). Vincent Scully finds the representation of both centralized and dispersed power in the Capitol building (Scully, *American Architecture and Urbanism*, 72).
26. Eugene F. Hemrick, *One Nation Under God: Religious Symbols, Quotes, and Images in Our Nation's Capital* (Huntington, IN: Our Sunday Visitor, 2001); Meyer, *Myths in Stone*, 81. Though the Lincoln Memorial's classical reference suggests Lincoln is placed in the role of a Greek god, the Parthenon was intentionally built to be more accessible to ordinary citizens, suggesting a democratization of hero worship and "humanized gods" (Gelertner, *A History of American Architecture*, 12).
27. Norma Evenson, "Monumental Spaces," in *The Mall in Washington, 1791–1991*, ed. Richard Longstreth (Washington, D.C.: National Gallery of Art, 1991), 33; Meyer, *Myths in Stone*, 8, 9; Richard Guy Wilson, "High Noon on the Mall: Modernism versus Traditionalism, 1910–1970," in *The Mall in Washington, 1791–1991*, ed. Richard Longstreth (Washington, D.C.: National Gallery of Art, 1991), 163. See also Abbott, *Political Terrain*, 115, 117–18, 131.
28. Pierson, *American Buildings*, 375. Jefferson called the Capitol building a "temple," albeit one "dedicated to the sovereignty of the people." Before being selected to plan the city, L'Enfant had suggested that it should express political power, writing that the seat of government should "give an idea of the greatness of the [American] empire" (Bowling, *The Creation of Washington, D.C.*, 6). But in the late eighteenth century, baroque city planning was seen as emblematic of beauty and rationalism rather than autocracy (Gutheim and Lee, *Worthy of the Nation*, 26–27), and the original plan was, in fact, less rigid and geometric than how it was later implemented, as we discuss below.
29. Evenson, "Monumental Spaces," 25–26. As another author puts it, "A great classical building will have a simple message for everyone while offering the knowledgeable spectator layer upon layer of more profound meaning." See Robert Adam, *Classical Architecture: A Comprehensive Handbook to the Tradition of*

Classical Style (New York: Harry N. Abrams, Inc., 1990), 1. Evenson notes the "symbolic adaptability of classicism" as shown in its appeal to Stalin and Hitler (Evenson, "Monumental Spaces," 27). Baroque is attractive for other reasons, too: for instance, it eases the flow of traffic and contributes to urban sanitation and health (Evenson, "Monumental Spaces," 21; Kostof, *The City Shaped*, 217).

30. Thomas, *The Lincoln Memorial*, 158–162; see also chapter 2. Architecture can also serve a symbolic purpose based on the timing of a building's construction. For instance, because the Capitol dome was completed during the Civil War, it became a symbol of the Republican Party's dedication to the Union (and optimism that the Union would prevail).
31. Tavernor, *Palladio and Palladianism*, 36–37; Vitruvius, *On Architecture*, book III, ch. 1. This symbolic meaning made classicism attractive to many Renaissance and Enlightenment architects and thinkers, though not necessarily Jefferson (Gelertner, *A History of American Architecture*, 39–40, 87, 98–99, Tavernor, *Palladio and Palladianism*, 37). One of the most important European architects who helped popularize neoclassicism in England and thus, by extension the United States was the Italian architect Palladio (Tavernor, *Palladio and Palladianism*).
32. Quoted in Jackson, "L'Enfant's Washington," 398. The city, however, was not literally in a swamp, as is commonly believed (Bowling, *The Creation of Washington, D.C.*, 237–238).
33. Dolley Madison's efforts to develop an attractive, politically oriented social scene in the city may have helped stop the plan. Catherine Allgor, *Parlor Politics: In Which the Ladies of Washington Help Build a City and a Government* (Charlottesville, VA: University of Virginia Press, 2000), 95–99; Catherine Allgor, *A Perfect Union: Dolley Madison and the Creation of the American Nation* (New York: Henry Holt and Company, 2006), 326–327. Talk of leaving the Potomac site for St. Louis, among other places, was seriously discussed after the Civil War; see chapter 9.
34. Charles Dickens, *American Notes for General Circulation* (New York: St. Martin's Press, 1985 [1842]). For more descriptions of the disappointing condition of Washington in the early to mid-1800s, see Reps, *Monumental Washington*, 38–42. Congress's jurisdiction over the city could be an impediment to Washington's development, which sometimes led city advocates to urge that Congress surrender its control of all or part of the District of Columbia. Bowling, *The Creation of Washington, D.C.*, 240–241; Jon A. Peterson, "The Senate Park Commission Plan for Washington, D.C.: A New Vision for the Capital and the Nation," in *Designing the Nation's Capital: The 1901 Plan for Washington, D.C.*, eds. Sue Kohler and Pamela Scott (Washington, D.C.: U.S. Commission of Fine Arts, 2006), 4.
35. Gelernter, 1999, 132–134; Moeller, 2006. Robert Pierson calls the Treasury Building "the most accomplished work of the rational phase of American Neoclassicism" (Pierson, 1970, 417).
36. Abbott, *Political Terrain*, 53–54; O'Malley, "A Public Museum of Trees," 61–62; Reps, *Monumental Washington*, 50.

37. Some legislators protested against the building's planned obstruction of the view, but in vain; Constance McLaughlin Green, *Washington, Vol. 1: Village and Capital, 1800–1878* (Princeton, NJ: Princeton University Press, 1962), 136–137.
38. Berg, *Grand Avenues*, 186–87, 190; Jennings, "Artistry as Design," 267, 274–275; O'Malley, "A Public Museum of Trees"; Peets, *On the Art of Designing Cities*, 16–17, 29–34, 50, 81; Reps, *Monumental Washington*, 42–44, 50–53; see also chapter 2. Though the off-axis monument maintained the White House's vista of the Potomac River, it prevented the monument from being framed by nearly any possible street in the city (Peets, *On the Art of Designing Cities*, 50, 81). Most agree that the monument's placement was due to insecure ground at the original proposed site, though Pamela Scott argues that it was done to fit the English-style garden of the Mall and to avoid disturbing Jefferson's meridian stone (Scott, "This Vast Empire," 51–53). Some potential deviations from L'Enfant's map, such as a university on the Mall, were avoided (Reps, *Monumental Washington*, 36–37).
39. Peterson, "The Senate Park Commission Plan," 4; Reps, *Monumental Washington*, 58–59; Donald Press, "South of the Avenue: From Murder Bay to the Federal Triangle," *Records of the Historical Society, Washington, D.C.* 51 (1984): 51–70; Kirk Savage, *Monument Wars: Washington, D.C., The National Mall, and the Transformation of the Memorial Landscape* (Berkeley: University of California Press, 2009), 13.
40. Gelertner, *A History of American Architecture*, 151–153, 162–163; James M. Goode, *Washington Sculpture: A Cultural History of Outdoor Sculpture in the Nation's Capital* (Baltimore: Johns Hopkins University Press, 2008); Gutheim and Lee, *Worthy of the Nation*, 80, 124–125; Tomas S. Hines, "The Imperial Mall: The City Beautiful Movement and the Washington Plan of 1901–1902," in *The Mall in Washington, 1791–1991*, ed. Richard Longstreth (Washington, D.C.: National Gallery of Art, 1991), 85–86; Lethbridge, "The Architecture of Washington, D.C.," 8–9, 11–12; Moeller, *AIA Guide to the Architecture of Washington, D.C.*; *New York Times*, November 6, 1898; Reps, *Monumental Washington*, 61, 66; Marcus Whiffen and Frederick Koeper, *American Architecture, Volume I: 1607–1860* (Cambridge, MA: The M.I.T. Press, 1981), 207. Egyptian Revival could be considered a variant of the Greek Revival (Whiffen and Koeper, *American Architecture*, 178). For other developments that did not conform to L'Enfant's vision, see Jennings, "Artistry as Design" and Peets, *On the Art of Designing Cities*.
41. The catalyst for a new height law was more practical than artistic: people feared that the Cairo building would overshadow its neighbors and that its upper floors could not be reached by fire trucks. The building, ironically, was modeled after a building from the 1893 World's Fair, the event that marked the resurgence of neoclassicism in Washington and elsewhere. Congress enacted height restrictions for the city in 1899 and 1910. See James M. Goode, *Best Addresses: A Century of Washington's Distinguished Apartment Houses* (Washington, D.C.: Smithsonian Institution Press, 1988), xix, 14–18; Moeller, *AIA Guide to the Architecture of Washington, D.C.*, 259; U.S. House of Representatives, Committee on the

District of Columbia, *Building Height Limitations*, Serial No. S-5 (Washington, D.C.: Government Printing Office, 1976).

42. Gutheim and Lee, *Worthy of the Nation*, 126; Peterson, "The Senate Park Commission Plan," 6–14. Political compromises to get the measure passed prevented the commission from being empowered to do more, though it ended up doing so anyway.
43. Hines, "The Imperial Mall," 81–85; Peterson, "The Senate Park Commission Plan," 14, 16. Peterson, "The Senate Park Commission Plan," and Reps, *Monumental Washington,* discuss the origins of the commission and the development of the McMillan Plan in detail.
44. Reps, *Monumental Washington*, chapter 5, 133; Peterson, "The Senate Park Commission Plan"; Savage, *Monument Wars*, 13–15. For more on Savage's concept of public space, see chapter 2.
45. Gelertner, *A History of American Architecture*, 202; Hines, "The Imperial Mall," 79–81, 96; Peterson, "The Senate Park Commission Plan"; Reps, *Monumental Washington*, 194–195; Wilson, "High Noon on the Mall," 151. Burnham similarly saw beauty, not symbolism, as the justification for long avenues, writing that "there is a true glory in mere length, in vistas longer than the eye can reach, in roads of arrow-like purpose that speed unswerving in their flight" (quoted in Kostof, *The City Shaped*, 235). For more on the City Beautiful movement, see William H. Wilson, *The City Beautiful Movement* (Baltimore: Johns Hopkins University Press, 1994).
46. Gutheim and Lee, *Worthy of the Nation*, 130–131, 138; Reps, *Monumental Washington*, 140–143.
47. Abbott, *Political Terrain*, 116–117; Moeller, *AIA Guide to the Architecture of Washington, D.C.*; Goode, *Best Addresses*, 25; Gutheim and Lee, *Worthy of the Nation*, 156; Savage, *Monument Wars*, 167–169.
48. Cynthia R. Field and Jeffrey T. Tilman, "Creating a Model for the National Mall: The Design of the National Museum of Natural History," *Journal of the Society of Architectural Historians* 63, no. 1 (2004): 52–73; Peterson, "The Senate Park Commission Plan," 35–36; Reps, *Monumental Washington*, 144–149. The museum redesigned at the last minute is the current Smithsonian Museum of Natural History.
49. Abbott, *Political Terrain*, 164; Gutheim and Lee, *Worthy of the Nation*, 138–139, 157, 167, 178–179, 193–194, 223–224, 331, 348–352; Reps, *Monumental Washington*, 154, 196–197; David C. Streatfield, "The Olmsteads and the Landscape of the Mall," in *The Mall in Washington, 1791–1991*, ed. Richard Longstreth (Washington, D.C.: National Gallery of Art, 1991), 130–136; Thomas, *The Lincoln Memorial*, chapter 3; Wilson, "High Noon on the Mall." The National Capital Park and Planning Commission was restructured again as the National Capital Planning Commission in 1952, while the Commission of Fine Arts was given still greater responsibilities by Congress in 1930 (Gutheim and Lee, *Worthy of the Nation*, 208, 259). The leadership of several key members of these organizations, including Charles Moore and Frederick

Law Olmsted Jr., as well as Glenn Brown, ensured their influence, especially on the development of the Mall (Witold Rybczynski, "'A Simple Space of Turf': Frederick Law Olmstead Jr.'s Idea for the Mall," in *The National Mall: Rethinking Washington's Monumental Core*, eds. Nathan Glazer and Cynthia R. Field (Baltimore: Johns Hopkins University Press, 2008); Streatfield, "The Olmsteads," 130; Thomas, *The Lincoln Memorial*, 80; Wilson, "High Noon on the Mall," 154–155).

50. This includes planting formal gardens around the Washington Monument, putting the monument in line with the White House and Capitol, and building monumental government office buildings around the White House (Gutheim and Lee, *Worthy of the Nation*, 135–136, 209; Peets, *On the Art of Designing Cities*, 61–68, 72–73; Reps, *Monumental Washington*, 176–179).
51. Edmond N. Bacon, Design of Cities (New York: Viking Press, 1967), 208; Hines, "The Imperial Mall," 92; Mumford, *The City in History*, 406; Peets, *On the Art of Designing Cities*, 27, 70; Reps, *Monumental Washington*, 136–137. See also Jane Jacobs, *The Death and Life of Great American Cities* (New York: Random House, 1961), 24–25, 173. For imaginings of what L'Enfant's city might look like today had his proposal been fully implemented, see Jackson, "L'Enfant's Washington" and Peets, *On the Art of Designing Cities*.
52. Evenson, "Monumental Spaces," 33; Gutheim and Lee, *Worthy of the Nation*, 181; Rybczynski, "A Simple Space of Turf," 63; Zachary Schrag, *The Great Society Subway: A History of the Washington Metro* (Baltimore: Johns Hopkins University Press, 2006); Wilson, "High Noon on the Mall," 145. Belief that the Mall is national, not local, predates the twentieth century; see e.g. O'Malley, "A Public Museum of Trees," 65.
53. Gutheim and Lee, *Worthy of the Nation*, 136; Hines, "The Imperial Mall," 95; Kostof, *The City Assembled*, 280; Mumford, *The City in History*, 387. See also Kostof, *The City Assembled*, 266–279. For criticisms of modernist buildings in Washington, see Carl Feiss, 1975, "Washington, D.C.: Symbol and City," in *World Capitals: Toward Guided Urbanization*, ed. Hanford Wentworth Eldredge (New York: Anchor Press, 1975); Lethbridge, "The Architecture of Washington, D.C.," 14, 16.
54. Constance McLaughlin Green, *Washington, Vol. 2: Capital City, 1879–1950* (Princeton, NJ: Princeton University Press, 1962), 16. The CFA, discussed earlier, played an important part in the process of determining the design of Metro's subway systems (Schrag, *The Great Society Subway*, 83–93).
55. Margaret E. Farrar, *Building the Body Politic: Power and Urban Space in Washington, D.C.* (Urbana: University of Illinois Press, 2008), 15; Nathan Glazer, "Monuments, Modernism, and the Mall," in Glazer and Field, *The National Mall*, 126; Kostof, *The City Shaped*, 215. Congress did not require development outside the preplanned area of Washington to follow the L'Enfant plan until 1888. The District of Columbia was considered distinct from the city of Washington until Congress unified them in 1871.

Chapter 2

1. For more on the religious symbolism of Washington, see chapter 1.
2. Michael Lewis, "The Idea of the American Mall," in *The National Mall: Rethinking Washington's Monumental Core*, eds. Nathan Glazer and Cynthia R. Field (Baltimore: Johns Hopkins University Press, 2008), 1–25; G. Martin Moeller, Jr., *Guide to the Architecture of Washington* (Baltimore: Johns Hopkins University Press, 2006), 69–71; Kirk Savage, *Monument Wars: Washington D.C., the National Mall, and the Transformation of the Memorial Landscape* (Berkeley: University of California Press, 2009). According to the National Park Service, the Mall receives around 24 million visitors per year. See "National Park Service FAQ," http://www.nps.gov/mall/faqs.htm, and Michael Bednar, *L'Enfant's Legacy; Public Open Spaces in Washington, D.C.* (Baltimore: John's Hopkins University Press, 2006).
3. *The American Heritage Dictionary of the English Language*, third edition, s.v. "Memorial" and "Monument."
4. Linton Weeks, "Maya Lin's 'Clear Vision,'" *The Washington Post*, October 20, 1995; Savage, *Monument Wars*, 266; Nicolaus Mills, *Their Last Battle; The Fight for the National World War II Memorial* (New York: Basic Books, 2004), 51; John Bodnar, *Remaking America; Public Memory, Commemoration, and Patriotism in the Twentieth Century* (Princeton: Princeton University Press, 1992), 6. See also James Goode, *Washington Sculpture; A Cultural History of Outdoor Sculpture in the Nation's Capital* (Baltimore: Johns Hopkins University Press, 2008).
5. Erika Doss, *Memorial Mania; Public Feeling in America* (Chicago: University of Chicago Press, 2010), 37–41; Nathan Glazer, "Monuments, Modernism, and the Mall" in *The National Mall: Rethinking Washington's Monumental Core*, eds. Nathan Glazer and Cynthia R. Field. (Baltimore: Johns Hopkins University Press, 2008), 117–133. The National Park Service has its own nomenclature for parks falling into the national monument category: they preserve "at least one nationally significant resource." Devils Tower National Monument in Wyoming, placed in this category by the U.S. government in 1906, was the first of such monuments. "A National Monument, Memorial, Park . . . What's the Difference?" National Park Service, http://www.nationalatlas.gov/articles/government/a_nationalparks.html.
6. Doss, *Memorial Mania*, 13–15. "Memorial Mania" is not the focus of this chapter, but Doss defines it as "an obsession with issues of memory and history and an urgent desire to express and claim those issues in visibly public contexts." For Doss, memorials "are bodies of feeling, cultural entities whose social, cultural, and political meaning are determined by the emotional states and needs of their audiences." Doss, *Memorial Mania*, 2, 46.
7. Bodnar, *Remaking America*, 13–14.
8. James Sterling Young, *The Washington Community, 1800–1828* (New York: Columbia University Press, 1966), 41–48; Thomas Moore, "To Thomas Hume, From the City of Washington," in *The Poetical Works of Thomas Moore with Life* (Edinburgh: Gall and Inglis, 1859), 265.

9. Young, *Washington Community*, 41. Benjamin Henry Latrobe's proposal for a national university presented an intriguing opportunity for the establishment of a well-designed cultural institution in or near the center of the city. Latrobe's plan, inspired by Thomas Jefferson's University of Virginia and in keeping within the tradition of L'Enfant's own vision, would have given the area a formality and elegance it had lacked. Though favored by George Washington, it gained no financial support when it was proposed in 1816. Lewis, "The Idea of the American Mall," 15–16, 21.
10. Jon A. Peterson, "The Senate Park Commission Plan for Washington, D.C.: A New Vision for the Capital and the Nation" in *Designing the Nation's Capital: The 1901 Plan for Washington, D.C.* (Washington, D.C.: U.S. Commission on Fine Arts, 2006), 1–47; Lewis, "The Idea of the American Mall," 11–25; quotes from Mills, *Their Last Battle,* xxviii. See also chapter 1.
11. Savage, *Monument Wars*, 13, 195–196, 210–211.
12. Savage, *Monument Wars*, 78, 211. Apparently the equestrian statue form was particularly suited, as this list suggests, to the celebration of the male military hero. In 1922, however, the Société de Femmes de France presented an equestrian statue of Joan of Arc to the women of the United States, on behalf of the women of France. Joan of Arc is the only woman to be memorialized in an equestrian statue in Washington, D.C. Allan M. Heller, *Monuments and Memorials of Washington, D.C.* (Atglen, PA: Schiffer Publishing, Ltd., 2006), 181.
13. Michael Lewis details the ways that L'Enfant's vision for the Mall drew from Thomas Jefferson's earlier designs for the city. See Lewis, "Idea of the American Mall," 13–14; Savage, *Monument Wars*, 79–80, 195–197, quotes on p. 196; Heller, *Monuments and Memorials*, 188–190.
14. As Lewis notes, Downing was the first to "reconcile the three different roles that the Mall had been asked to perform: the recreational, the didactic, and the symbolic." Lewis, "The Idea of the American Mall," 21–22.
15. Mills, *Their Last Battle*, 49–50; Heller, *Monuments and Memorials*, 121.
16. Quoted in Heller, *Monuments and Memorials*, 121.
17. Heller, *Monuments and Memorials*, 121.
18. Savage, *Monument Wars,* 13–14, 19, 147.
19. Savage, *Monument Wars*, 147–148, 152.
20. Lewis, "The Idea of the American Mall," 23.
21. Quoted in Kenneth R. Fletcher, "A Brief History of Pierre L'Enfant and Washington, D.C.," *Smithsonian Magazine*, May 1, 2008.
22. Savage, *Monument Wars*, 25. On L'Enfant, see Scott W. Berg, *Grand Avenues: The Story of the French Visionary Who Designed Washington, D.C.* (New York: Pantheon Books, 2007).
23. Berg, *Grand Avenues*, passim; Heller, *Monuments and Memorials*, 27.
24. Another memorial, the Emancipation Monument in Judiciary Square created by Thomas Ball, dedicated more than four decades earlier, also features a statue of Lincoln. See Heller, *Monuments and Memorials*, 173–175.

25. The Grant Memorial, now below the western side of the Capitol, was initially planned for the current Lincoln Memorial location. Peterson, "The Senate Park Commission Plan," 26.
26. Christopher Thomas, *The Lincoln Memorial and American Life* (Princeton: Princeton University Press, 2002), 55.
27. Thomas, *The Lincoln Memorial and American Life*, 56. The criteria summarize Thomas's excellent treatment of the design process.
28. Thomas, *The Lincoln Memorial and American Life*, 55; Savage, *Monument Wars*, chapter 4.
29. In 1923, the year after the Lincoln Memorial's dedication, about a third to a half million visits were made to the structure. See Thomas, *The Lincoln Memorial and American Life*, 146.
30. Thomas, *The Lincoln Memorial and American Life*, 144. Here, Thomas borrows from the work of Pierre Nora, who discusses the Lincoln Memorial as a "memory site." See Pierre Nora, "Between Memory and History; Les Lieux de Mémoire" *Representations*, 26 (Spring 1989).
31. Thomas, *The Lincoln Memorial and American Life*, 158; Bodnar, *Remaking America,* 13–14.
32. Thomas, *The Lincoln Memorial and American Life*, 146–147, 150–151.
33. Quoted in Thomas, *The Lincoln Memorial and American Life,* 153.
34. See also chapter 8.
35. Scott Sandage, "A Marble House Divided: The Lincoln Memorial, the Civil Rights Movement, and the Politics of Memory, 1939–1963," *Journal of American History*, vol. 80, no. 1 (June 1993), 135–167, 138, 152. See also chapter 5.
36. Sandage, "A Marble House Divided," 138; Savage, *Monument Wars*, 257.
37. Bodnar, *Remaking America*, 3–6. In 1993 the Vietnam Women's Memorial was added to the grounds.
38. See especially Thomas B. Allen, *Offerings at the Wall: Artifacts from the Vietnam Veterans Memorial Collection* (Atlanta: Turner Publishing, 1995).
39. Savage, *Monument Wars,* 265–266.
40. Blair Ruble, *Washington's U Street, A Biography* (Woodrow Wilson Center Press, The Johns Hopkins University Press: Washington, D.C. and Baltimore, 2010), 283–284.
41. Doss, *Memorial Mania*, 187–192.
42. Doss, *Memorial Mania*, 197.
43. Mills, *Their Last Battle,* 164–165; Savage, *Monument Wars*, 298.
44. Edward Rothstein, "A Mirror of Greatness, Blurred," *The New York Times*, August 25, 2011.

Chapter 3

1. According to *Travel and Leisure,* three of the ten most visited museums in the world in 2011 were located in Washington, D.C.: the National Air and Space Museum, the National Museum of Natural History, and the National Gallery of

Art. Lyndsey Matthews, "World's Most-Visited Museums," *Travel and Leisure*, November 2011, http://www.travelandleisure.com/articles/worlds-most-visited-museums/1.

2. For a list of all of the historic house sites that function as museums and tourist attractions in the greater metropolitan area, see *Historic House Museum Consortium*, http://dchousemuseums.org/.
3. For more about the changing role of visitors in the museum experience, see Graham Black, *The Engaging Museum: Developing Museums for Visitor Involvement* (New York: Routledge, 2005), 7–74.
4. For more about this concept, see chapter 6 in Nina Simon, *The Participatory Museum* (Santa Cruz, CA: Museum 2.0, 2010). Simon, an advocate of such inclusion of visitors in the process of exhibit creation, argues that visitors' voices can inform and invigorate project designs and public programs in museums.
5. Accessible through cell phones or MP3 players that visitors can borrow from the museum, this creative response to the challenge of how to provide interpretive information about the art while not overcrowding the space reflects the Luce Center's ability to evaluate their exhibitions and continual desire to improve the visitor experience.
6. "All visitors to the Luce Foundation Center discover something that captures their interest . . . In conceiving the center, we wanted to provide visitors with a genuine behind-the-scenes experience—an opportunity to walk through art storage and see thousands of artworks that are normally kept off-site." Georgina Bath, "Visible Storage at the Smithsonian American Art Museum," in *The Manual of Museum Management,* second edition. Eds. Gail Dexter Lord and Barry Lord (Lanham, MD: AltaMira Press, 2009), 180.
7. *Museums, Libraries and 21st Century Skills* (Institute of Museum and Library Services: Washington, D.C., 2009).
8. Online visitors are invited to upload pictures of their faces and then go through the steps to create the scene. When complete, individuals can include their scene in an online gallery and are given the option to e-mail the image as well. *Lincoln's Cottage at Soldiers' Home*, http://www.lincolncottage.org.
9. See chapter 1.
10. According to the National Capital Planning Commission, all of the sites for museums on the National Mall had been committed. See National Capital Planning Commission, "Extending the Legacy: Planning America's Capital for the 21st Century 1997," http://www.ncpc.gov/DocumentDepot/Publications/Legacy/Legacy_Introduction.pdf. City planners have identified numerous potential locations throughout the city for new museums that would connect with long-term development plans for the area. See National Capital Planning Commission, "Memorials and Museums Master Plan 2001. Updated 2006," http://www.ncpc.gov/ncpc/Main(T2)/Planning(Tr2)/2MPlan.html and National Capital Planning Commission, "Monumental Core Framework Plan 2009," http://www.ncpc.gov/ncpc/Main(T2)/Planning(Tr2)/FrameworkPlan.html.

11. For more information about the history of the Smithsonian Institution, see Heather Ewing, *The Lost World of James Smithson: Science, Revolution, and the Birth of the Smithsonian* (New York: Bloomsbury, 2007) and Nina Burleigh, *The Stranger and the Statesman: James Smithson, John Quincy Adams, and the Making of America's Greatest Museum: The Smithsonian* (New York: HarperCollins, 2004).
12. See *About the National Gallery of Art*, http://www.nga.gov/content/ngaweb/about.html and *History of the United States Holocaust Memorial Museum*, http://www.ushmm.org/wlc/en/article.php?ModuleId=10005782.
13. Jacqueline Trescott, "Ant-Covered Jesus Video Removed from Smithsonian after Catholic League Complaints," *Washington Post*, December 1, 2010.
14. Heather Ewing and Amy Ballard, *A Guide to Smithsonian Architecture* (Washington, D.C.: Smithsonian Books, 2009), 30, 40; *United States National Museum*, http://siarchives.si.edu/history/exhibits/arts/usnm.htm.
15. Catherine Lewis, *The Changing Face of Public History: The Chicago Historical Society and the Transformation of An American Museum* (DeKalb: Northern Illinois University Press, 2005), 7–8; Steven Lubar and Kathleen M. Kendrick, *Legacies: Collecting America's History at the Smithsonian* (Washington, D.C.: Smithsonian Institution Press, 2001), 206.
16. The famous 1969–1970 rodent exhibit, designed by museum director John Kinard, was called *The Rat: Man's Invited Affliction*. See Edmund Barry Gaither, "'Hey! That's Mine': Thoughts on Pluralism and American Museums," in *Reinventing the Museum: Historical and Contemporary Perspectives on the Paradigm Shift*, ed. Gail Anderson (Walnut Creek, CA: AltaMira Press, 2004), 113–114.
17. Portia James, "Building a Community-Based Identity at Anacostia Museum," in *Heritage, Museums and Galleries: An Introductory Reader*, ed. Gerard Corsane (New York: Routledge, 2005). For more information about the Anacostia Community Museum, see Smithsonian Institution Archives, "Anacostia Community Museum," http://siarchives.si.edu/history/anacostia-community-museum and The Anacostia Community Museum, http://anacostia.si.edu/.
18. Amanda J. Cobb, "The National Museum of the American Indian: Sharing the Gift," *American Indian Quarterly* 29, nos. 3 and 4 (Summer and Fall 2005): 375; Ewing and Ballard, *A Guide to Smithsonian Architecture*, 137; Julia M. Klein, "Two New History Museums Put Their Ideals on Display," *Chronicle of Higher Education* 51, no. 6 (October 1, 2004): B15–16. An example of this is in the Our Lives exhibition, where visitors are invited to play the traditional game of peon with four natives on the screen. The visitor learns the rules of the game and then chooses each move by using a touch screen. After each choice, the players on the screen respond with their move. This "choose your own adventure" approach allows the visitor to feel as if he is actually playing the game.
19. As Amanda J. Cobbs observes, "Our Universes, Our Peoples, and Our Lives may initially be difficult for visitors to interpret because they cannot be viewed or 'read' in the usual way." See Cobb, "Sharing the Gift," 374.
20. The groundbreaking ceremony for the museum provided an opportunity for a variety of museum staff and government officials who have been involved in making

this museum a reality to discuss the ideals they are striving for in the museum. For more, see Marisol Bello, "Obama, Laura Bush Break Ground for African American Museum," *USA Today*, February 22, 2012, and Patricia Cohen, "Making Way for a Dream in the Nation's Capital," *New York Times*, February 22, 2012.

21. *National Museum of African American History and Culture*, http://nmaahc.si.edu/.
22. See Kate Taylor, "National Latino Museum Plan Faces Fight," *The New York Times*, April 20, 2011 and *National Women's History Museum*, http://www.nwhm.org/.
23. Andrée Gendreau, "Museums and Media: A View from Canada," *The Public Historian* 31, no. 1 (Winter 2009): 35–45.
24. *President Lincoln's Cottage at the Soldiers' Home*, http://lincolncottage.org/.
25. For more information about the *Enola Gay* controversy, see Edward T. Linenthal's "Anatomy of a Controversy" and Richard H. Kohn's "History at Risk: The Case of the *Enola Gay*" in Edward T. Linenthal and Tom Engelhardt, eds, *History Wars: The Enola Gay and Other Battles for the American Past* (New York: Henry Holt and Company, 1996), 9–62, 140–170.
26. For a detailed history of the process of creating this museum, see Edward Linenthal, *Preserving Memory: The Struggle to Create America's Holocaust Museum* (New York: Columbia University Press, 2001).
27. One aspect of the Carnegie Library Building that made it particularly attractive for the museum venture was the significance attached to the fact the space had never been segregated when it functioned as the Central Public Library. DC Cultural Tourism, "Central Public Library," *African American Heritage Trail*, http://www.culturaltourismdc.org/things-do-see/central-public-library-african-american-heritage-trail.
28. Jacqueline Trescott, "City Museum to Close Its Galleries; Troubled Venue Will Continue to Rent Space for Events," *The Washington Post*, October 9, 2004, A01.
29. Arcynta Ali Childs, "Tour the American History Museum with an American Girl," March 1, 2011, *Smithsonian.com*, http://blogs.smithsonianmag.com/aroundthemall/2011/03/tour-the-american-history-museum-with-an-american-girl/; J. Gruber, "Smithsonian Remakes Its Transportation Exhibit: Corporate Sponsors Help with First Major Change in Four Decades," *Trains* 63, no. 3 (March 2003): 76–77; Robert L. Jackson, "Smithsonian's New Ocean Planet Exhibit Catches a Wave: Corporate Sponsorship," *The Baltimore Sun*, June 7, 1995; I. Michael Heyman, "Museums and Marketing," January 1998, *Smithsonian Magazine*, http://www.smithsonianmag.com/history-archaeology/heyman_jan98-abstract.html.
30. See *U.S. Capitol Visitor Center*, http://www.visitthecapitol.gov/exhibitions/index.html; *Library of Congress*, http://www.loc.gov/exhibits/; *National Archives*, http://www.archives.gov/nae/visit/gallery.html.
31. See *Washington, D.C. Trails and Tours*, http://www.culturaltourismdc.org/things-do-see/trails-tours.
32. Such an addition to the highly limited space on the Mall, however, has been fraught with controversy. Petula Dvorak, "Vietnam Wall Visitor Center Approved," *Washington Post*, August 4, 2006, and Meredith Somers, "Fight Goes on over Mall Visitors Center," *The Washington Times*, July 17, 2012.

Chapter 4

1. See chapter 9.
2. For more on this "Lockean liberal" tradition, see Louis Hartz, *The Liberal Tradition in America* (New York: Harcourt, Brace, & World, 1955).
3. Alexis de Tocqueville, *Democracy in America*, volume one, J. P. Mayer edition (New York: Harper and Row, 1969), 278; Kenneth R. Bowling, "From 'Federal Town' to 'National Capital': Ulysses S. Grant and the Reconstruction of Washington, D.C.," *Washington History* 14, no.1 (Spring/ Summer 2002): 10–14; Young, *The Washington Community*, chapter 3.
4. Abbott, *Political Terrain*, 128; Google Ngram Viewer, search term *inside the Beltway,* accessed September 11, 2011. On American's suspicion of cities, see Dennis R. Judd and Todd Swanstrom, *City Politics: The Political Economy of Urban America*, fifth edition (New York: Pearson/ Longman, 2006), 6–7.
5. See, e.g., Richard F. Fenno, Jr., *Home Style: House Members in Their Districts* (New York: HarperCollins Publishing, 1978), 97.
6. Lee Davidson, "Chaffetz Giving 'Cot-Side' Chats," *Deseret News*, February 27, 2009, http://www.deseretnews.com/article/705287666/Chaffetz-giving-cot-side-chats.html, accessed 8/23/11; David A. Fahrenthold, "The House Is Their Home," *Washington Post*, February 15, 2011; Sheryl Gay Stolberg, "Daschle, Democratic Senate Leader, Is Beaten," *New York Times,* November 3, 2004. For a recent example, see Shira Toeplitz, "Ex-Rep. Halvorson May Challenge Jackson," *Roll Call*, September 6, 2011. Some also choose to live in their offices for economic reasons: Washington, D.C. housing can be quite expensive, particularly if one is also maintaining a home in one's state or district. This was the explanation given by Congressman Trey Gowdy (R-SC) for not living in Washington, even when he chaired the House subcommittee with jurisdiction over the city (Mike DeBonis, "D.C. Keeps Wary Eye on House GOP," *Washington Post*, January 28, 2011).
7. Meyer, *Myths in Stone*.
8. Henry Adams, *Democracy* (New York: Harmony Books, 1981 [1880]), 9–10.
9. Ibid., 15.
10. Morris Fiorina, *Congress: Keystone of the Washington Establishment*, revised edition (New Haven: Yale University Press, 1989); Andrew King, "The Vulnerable American Politician," *British Journal of Political Science* 27 (1997): 1–22; David Mayhew, *Congress: The Electoral Connection* (New Haven: Yale University Press, 1974).
11. Some offices disproportionately display decorations featuring the lawmaker herself. Two reporters even constructed a "vanity index" of senators based on the amount of self-promoting office décor they featured (Darren Garnick and Ilya Mirman, "If the Walls Could Talk," *Slate,* October 29, 2010, http://www.slate.com/id/2272482).
12. Fenno, *Home Style*.
13. In 2006, for instance, Congressman Patrick Kennedy (D-RI), under the influence of prescription drugs, crashed his car near the Capitol but was merely

escorted home by Capitol police officers. For more examples from D.C.'s history, see Jon Katz, "2 Students Say Albert Was 'Drunk,'" *Washington Post*, September 13, 1972, and Drew Pearson and Robert S. Allen, "Boiled Bosoms" in Katharine Graham, *Katharine Graham's Washington*, edited by Katharine Graham (New York: Vintage Books, 2002), 205–208.

14. Marc Fisher, "Chevy Chase, 1916: For Everyman, a New Lot in Life," *Washington Post*, February 15, 1999; Jackson, *Crabgrass Frontier*, 123. Others, such as local banker Charles Glover, were also instrumental in the creation of the park. Many were motivated by altruistic considerations, including expanding urban recreation space and water supplies. See William B. Bushong, "Glenn Brown and the Planning of the Rock Creek Valley," *Washington History* 14, no. 1 (2002): 56–71 and Barry Mackintosh, *Rock Creek Park: An Administrative History* (Washington, D.C.: National Park Service, 1985), 2–15. Stewart also used advance knowledge about planned city development in the early 1870s to profit from investments in soon-to-be-developed areas in western Washington. See Kathryn Allamong Jacob, "'Like Moths to a Candle': The Nouveaux Riches Flock to Washington, 1870–1900" in *Urban Odyssey: A Multicultural History of Washington, D.C.*, ed. Francine Curro Cary. (Washington, D.C.: Smithsonian Institution Press, 1996), 83–84. Another legislator who invested heavily in Washington's growth was Senator John Sherman of Ohio, who helped create LeDroit Park, an early Washington suburb, and the Columbia Heights neighborhood; he also introduced the bill in the Senate that would eventually create Rock Creek Park. See Kathryn Allamong Jacob, *Capital Elites: High Society in Washington, D.C. after the Civil War* (Washington, D.C.: Smithsonian Institution Press, 1995), 163–164; Mackintosh, *Rock Creek Park.*
15. Charles Wesley Harris, *Congress and the Governance of the Nation's Capital: The Conflict of Federal and Local Interests* (Washington, D.C.: Georgetown University Press, 1995), 152; Sue Anne Pressley Montes, "Report Adds to Debate over Putting Meters in D.C. Cabs," *Washington Post*, July 28, 2007; Schrag, *The Great Society Subway*, 314. This congressional ban on meters dated back at least to the early 1930s; see, for example, "A 20-Cent Attitude," *Washington Post,* July 7, 1935, and "Cab Meters Rule Upheld in Reply to House Attack," *Washington Post,* January 7, 1932. Congress's power over the District of Columbia is discussed in more detail in chapter 7.
16. Some national politicians (both acting and retired) and their spouses have tried to improve the city for generally unselfish reasons. For example, Amos Kendall, former postmaster general under Presidents Andrew Jackson and Martin Van Buren, successfully lobbied Congress to fund the establishment of the city's famous school for the deaf, Gallaudet University (Green, *Washington*, vol. 1, 219–220). Dolley Madison was another such early figure; see Catherine Allgor, *A Perfect Union: Dolley Madison and the Creation of the American Nation* (New York: Henry Holt & Co., 2006), 324–325. More recently, First Lady Michelle Obama participated in volunteer work around the city, including food kitchens, a community health center, and Toys for Tots (Krissah Thompson, "Michelle Obama's Washington," *Washington Post*, September 26, 2013).

17. Samuel Kernell, *Going Public: New Strategies of Presidential Leadership*, 4th edition (Washington, D.C.: CQ Press, 2006); Richard Neustadt, *Presidential Power and the Modern Presidents: The Politics of Leadership from Roosevelt to Reagan* (New York: Free Press, 1991), 11; Matthew Soberg Shugart and John M. Carey, *Presidents and Assemblies: Constitutional Design and Electoral Dynamics* (New York: Cambridge University Press, 1992); Hedrick Smith, *The Power Game: How Washington Works* (New York: Ballantine Books, 1988), 13–14, quote p. 14; see also Matt Bai, "The Insiders," *New York Times Magazine*, June 7, 2009. For skeptical views of the belief that the mandate exists, see Robert A. Dahl, "The Myth of the Presidential Mandate," *Political Science Quarterly* 105 (1990): 3, 355–372. The bully pulpit tends to work far less well than people assume; see George C. Edwards III, *On Deaf Ears: The Limits of the Bully Pulpit* (New Haven: Yale University Press, 2003).
18. David Hagedorn, "Guess Who's Coming . . ." *Washington Post*, September 27, 2011; Smith, *The Power Game*, chapter 1.
19. Clifford Geertz, *Local Knowledge: Further Essays in Interpretive Anthropology* (New York: Basic Books, 1983), 124, 143; H. L. Mencken, *On Politics*, ed. Malcolm Moore (Baltimore: Johns Hopkins University Press, 1956), 249. It also echoes the "court-in-motion" of Moroccan kings who, by travelling throughout their kingdom, demonstrated their authority to more distant realms (Geertz, *Local Knowledge*, 136–138).
20. The high turnover is doubtless due in part to the relatively low pay that congressional aides earn. Over four-fifths of congressional staff are white and about evenly split by gender, though top staff are predominantly men: in 2010, nearly 70 percent of chiefs of staff and 60 percent of legislative directors were male. The median age of a House aide is thirty-five. Congressional Management Foundation, *House Staff Employment Study*, 2002, http://assets.sunlightfoundation.com.s3.amazonaws.com/policy/staff%20salary/2002%20House%20Staff%20employment%20Study.pdf, 2–3, 11, 15, 73, 75–76; U.S. House of Representatives, Chief Administrative Office, *2010 House Compensation Study* (Washington, D.C.: ICF International, 2010); Sunlight Foundation, "Keeping Congress Competent: Staff Pay, Turnover, and What It Means for Democracy," 2010, http://assets.sunlightfoundation.com.s3.amazonaws.com/policy/papers/Staff%20Pay%20Blogpost.pdf.
21. See, for example, Jacob, *Capital Elites*, 5, 25, and Jay Franklin, "Main Street-on-Potomac," in *Katharine Graham's Washington*, edited by Katharine Graham, 25. In the late nineteenth and early twentieth century, Washington's transient nature could also be attributed to the many wealthy who maintained winter homes in the city. The city's population has also felt less permanent at certain times in its history; Katharine Graham, for instance, wrote that during World War II, "with the buildup of the effort to prepare to fight the war, there was a sense of transience about the whole city" (Graham, *Katharine Graham's Washington*, 68). By 1900, defeated lawmakers were settling in Washington, D.C. in increasing numbers (Jacob, *Capital Elites*, 237).

22. David E. Lewis, *The Politics of Presidential Appointments* (Princeton, NJ: Princeton University Press, 2008), 22–23; Open Secrets, "Lobbying Database," 2011, http://www.opensecrets.org/lobby/index.php; Norman J. Ornstein, Thomas E. Mann, and Michael J. Malbin, *Vital Statistics on Congress 2008* (Washington, D.C.: Brookings Institution Press., 2008), 110; Pew Research Center, "The New Washington Press Corps," 2009, http://www.journalism.org/analysis_report/numbers; White House, *2011 Annual Report to Congress on White House Staff*, http://www.whitehouse.gov/briefing-room/disclosures/annual-records/2011. Another 300,000 people work for the executive branch in some capacity in the greater Washington metropolitan area, and an estimated 20,000 people intern in Washington every summer. U.S. Office of Personnel Management, "Table 2—Comparison of Total Civilian Employment of the Federal Government by Branch, Agency, and Area as of August 2009 and September 2009," September 2009, http://www.opm.gov/feddata/html/2009/September/table2.asp; Ross Perlin, "Five Myths about . . . Interns," *Washington Post*, May 20, 2011.
23. Shannon, "DC Mythbusting: No One Is From DC," *We Love DC* blog, September 15, 2011, http://www.welovedc.com/2009/09/15/dc-mythbusting-no-one-is-from-dc/.
24. See, e.g., Christopher Callahan, "Lagging Behind," *American Journalism Review* (August/September 2004), accessed August 23, 2011, http://www.ajr.org/article.asp?id=3738. Based on available personnel data, we estimate that 55 percent (251) of the 454 White House staff in 2010 were male (White House, *2011 Annual Report*). Though data on executive branch staff in D.C. is not readily available, over two-thirds of all U.S. executive branch employees in 2006 were white and 56 percent were men. U.S. Office of Personnel Management, "Table 1—Race/National Origin Distribution of Civilian Employment, Executive Branch Agencies, World Wide," September 2006, http://www.opm.gov/feddata/demograp/table1mw.pdf; U.S. Office of Personnel Management, "Table 1—Executive Branch (non-Postal) Employment by Gender, Race/National Origin, Disability Status, Veterans Status, Disabled Veterans," September 2006, http://www.opm.gov/feddata/demograp/table1-1.pdf.
25. Young, *The Washington Community*, chapters 4 and 5. In the early 1800s, a division developed between two competing social groups: those in the city for short-term political reasons and those who considered the city their permanent home. After the Civil War, the nouveau riche who saw Washington as an attractive wintering spot became a third social group. Jacob, *Capital Elites*, 56, 140–150; see also chapter 9.
26. J. Ross Eshleman, Barbara G. Cashion, and Laurence A. Basirico, *Sociology: An Introduction*, fourth edition (New York: HarperCollins College Publishers, 1993), 119, 123. On the nineteenth-century society of Washington, Jacob writes that "there were partisan, regional, and occupational cliques within official society. These sub-circles within official society often intersected." Jacob, *Capital Elites*, 77.
27. Ralph Linton, *The Study of Man: An Introduction* (New York: Appleton-Century-Crofts, 1936), 115. Wrote one journalist in 1884, Washington's society

"is not founded on wealth . . . [but] public station, temporarily conferred, whether directly or indirectly." Frank Oppel and Tony Meisel, eds., *Washington, D.C.: A Turn-of-the-Century Treasury* (Secausus, NJ: Castle, 1987), 129. But power is not the only basis for ranking in Washington society. For example, the city's national press corps ranked its members based on the length of their membership in their exclusive Gridiron Club (Joseph Alsop, "Dining-Out Washington," in Graham, *Katharine Graham's Washington*, 145). See also Mark Leibovich, *This Town: Two Parties and a Funeral—Plus Plenty of Valet Parking!—In America's Gilded Capital* (New York: Blue Rider Press, 2013), 23.

28. Graham, *Katharine Graham's Washington*, 451; Oppel and Meisel, *Washington, D.C.*, 432. On changes in the rank of diplomats relative to Washington politicians over time, see Hope Ridings Miller, *Embassy Row: The Life & Times of Diplomatic Washington* (New York: Holt, Rinehart and Winston, 1969), chapter 2.
29. Meg Greenfield, *Washington* (New York: PublicAffairs, 2001), 23; Smith, *The Power Game*, 97–101; Olive Ewing Clapper, *Washington Tapestry* (New York: McGraw-Hill, 1946), 165–167, quote p. 167. One popular guidebook from 1880s Washington identified over two dozen ranks among the city's officials alone; Frank G. Carpenter, *Carp's Washington* (New York: McGraw-Hill, 1960), 95. In 1960, the influential socialite Perle Mesta offered a more precise listing of rankings (Perle Mesta, "Bigwigs, Littlewigs, and No Wigs at All," in Graham, *Katharine Graham's Washington*, 167–168).
30. Graham, *Katharine Graham's Washington*, 687; Leibovich, *This Town,* 31; Talcott Parsons, *The Social System* (New York: The Free Press, 1951), chapter 9; "Lapel Pins to Identify Congressmen," *Youngstown Vindicator*, May 19, 1975; "Washington Wire," *Wall Street Journal,* May 16, 1975.
31. Eshleman, Cashion, and Basirico, *Sociology*, 128–129; Greenfield, *Washington*, 36; Leibovich, *This Town,* 92; Smith, *The Power Game*, 97.
32. Catherine Allgor, *Parlor Politics: In Which the Ladies of Washington Help Build a City and a Government* (Charlottesville, VA: University Press of Virginia, 2000), 5–6, 66, 87; Allgor, *A Perfect Union*, quotes pp. 191, 245; Fredrika J. Teute, "Roman Matron on the Banks of the Tiber River," in *A Republic For the Ages: The United States Capitol and the Political Culture of the Early Republic*, ed. Donald R. Kennon (Charlottesville, VA: University Press of Virginia, 1999); Young, *The Washington Community*, chapter 4.
33. Allgor, *Parlor Politics*, 120–124; Cynthia D. Earman, "Remembering the Ladies: Women, Etiquette, and Diversions in Washington City, 1800–1814," *Washington History* 12, no. 1 (Spring/ Summer 2000): 105–106. Formal dinners have remained an important activity for Washington elite, while social calls lasted as a tradition until roughly the 1930s, when "almost all of the attendant rigors of strict society had fallen by the wayside." Graham, *Katharine Graham's Washington*, 113.
34. Jacob, *Capital Elites*, 10. Jacqueline M. Moore, *Leading the Race: The Transformation of the Black Elite in the Nation's Capital, 1880–1920* (Charlottesville, VA: University Press of Virginia, 1999), 52–53; Burt Solomon, *The Washington Century: Three Families and the Shaping of the Nation's Capital* (New York: HarperCollins,

2004), 42; Thorstein Veblen, *The Theory of the Leisure Class* (New York: The MacMillan Company, 1899), 1, 75. For more, see chapter 8. For more on Cafritz, see Solomon, *The Washington Century*; the letter is recounted on p. 109.

35. DeNeen Brown, "The 'New' Washington Dinner Party," *Washington Post Magazine,* June 10, 2012. Descriptions of, and examples from, social life in Washington and how it changed during the nineteenth century can be found in Green, *Washington, Vol. 1*, 110–111, 155; Jacob, *Capital Elites*; and excerpts from Oppel and Meisel, *Washington, D.C.*, 88–91, 126–127, 131, and 333–351. The city's social elite, at least in its early decades, were not necessarily members of its political community. Joseph Alsop, for instance, noted that its members "were connected to one another not by their proximity to power (as is the case in what passes for social fashion in the capital today) or by fantastic wealth, but by a certain longevity, a modicum of breeding, and a decidedly southern sense of grand style." Alsop, "Dining-Out Washington," 140.
36. Trent Lott, *Herding Cats: A Life in Politics* (New York: HarperCollins, 2005), 250; see also Dan Zak, "Party Politics on Tap," *Washington Post*, June 6, 2011.
37. Paul Farhi, "Catering to the C-suite," *Washington Post,* April 26, 2013; Leibovich, *This Town,* 103; Roxanne Roberts and Amy Argetsinger, "Lee's Ripe for Roasting," *Washington Post*, February 11, 2011; Roxanne Roberts and Amy Argetsinger, "Spotting the Quasi (In)famous," *Washington Post*, February 14, 2011; Roxanne Roberts and Amy Argetsinger, "At Nerd Prom, the Glitzy Sets Grabs Control," *Washington Post*, May 2, 2011. For an older account of the Gridiron dinner, see Clapper, *Washington Tapestry*, 167–168.
38. Henri Tajfel and John Turner have been credited with introducing and developing the idea that social identity is central to understanding the formation of, and self-identification with, social groups, as well as the ability to contrast one's group with that of others. See, for example, Henri Tajfel and John C. Turner, "The Social Identity Theory of Inter-Group Behavior," in *Psychology of Intergroup Relations*, eds. Stephen Worchel and William G. Austin (Chicago: Nelson-Hall Publishers, 1986) and Turner et al., *Rediscovering the Social Group: A Self-Categorization Theory* (New York: Basil Blackwell, 1987).
39. Clapper, *Washington Tapestry*, 157.
40. Allgor, *A Perfect Union*, 61–62, 224–226, 278; Allgor, *Parlor Politics*, 24–27, 95–99; Earman, "Remembering the Ladies," 106–107; Smith, *The Power Broker*, 389–393, quote p. 389; Young, *The Washington Community*, 167–173, quote p. 168. Margaret Bayard Smith was also an important dispenser of patronage (Allgor, *Parlor Politics*, 132–144).
41. Greenfield, *Washington*; Katharine Graham, *Personal History* (New York: Vintage Books, 1997), 616; Smith, *The Power Game*, chapter 6.
42. Bowling, *The Creation of Washington, D.C.*, 241–245; Christopher Hayes, "Why Washington Doesn't Care about Jobs," *The Nation*, March 3, 2011; John Kelly, "Old D.C.'s Manifest Destiny: Staying Put," *Washington Post*, August 15, 2010; see also Thomas Frank, "The Bleakness Stakes," *Harper's Magazine*, November 2011 and chapter 9.

43. Allgor, *Parlor Politics*, chapter 5; Jacob, *Capital Elites*, 95–97, 104–106, quote p. 105.
44. Graham, *Personal History*, 610.
45. Patrick Gavin, "Brokaw Says 'No Thanks' to WHCD," *Politico's Guide to the White House Correspondents' Dinner*, April 26, 2013; Dana Milbank, "Journalists Gone Wild," *Washington Post*, May 1, 2011. See also Leibovich, *This Town*, ch. 6.

Chapter 5

1. Library of Congress, "Meet Me at the Willard: Famed Hotel Is Subject of Library Display," *Information Bulletin* 65 (2006): 90.
2. Lucy G. Barber, *Marching on Washington: The Forging of an American Political Tradition* (Berkeley, CA: University of California Press, 2002).
3. Steven L. Danver, *Revolts, Protests, Demonstrations, and Rebellions in American History: An Encyclopedia* (Santa Barbara, CA: ABC-CLIO LLC, 2011), i.
4. Barber, *Marching on Washington,* 7.
5. Barber, *Marching on Washington*, 11.
6. Michael S. Sweeney, "'The Desire for the Sensational': Coxey's Army and the Argus-Eyed Demons of Hell," *Journalism History* 23 (1997): 114.
7. Sweeney, "The Desire for the Sensational," 116.
8. The period following the Civil War through the beginning of the twentieth century was one of "rapidly developing social problems, but there was not much recognition of how serious the problems were nor was there any general acceptance of social responsibility for their solution." Phillip R. Popple and Leslie Leighninger, *Social Work, Social Welfare, and American Society,* eighth edition (Upper Saddle River, NJ: Pearson, 2011), 292. On the one hand, there was the perception that poverty was a necessary part of the human condition because it gave those who were not poor the opportunity to practice the virtue of charity. On the other hand there was the view that America was the land of plenty, and therefore poverty was the result of laziness, intemperance, or impiety. The two perspectives colored approaches to resolving the issue of poverty, even among early practitioners of the profession that came to be called social work, such as Jane Addams, the Nobel prize–winning woman credited with the beginning of the settlement house movement in the United States. See Louise W. Knight, "Changing My Mind: An Encounter with Jane Addams," *Affilia: Journal of Women & Social Work* 21(2006), 99.
9. Sweeney, "The Desire for the Sensational," 115.
10. Sweeney, "The Desire for the Sensational," 117.
11. Sweeney, "The Desire for the Sensational," 118.
12. "Coxey's Army Starts," *Washington Post,* March 26, 1894.
13. Sweeney, "The Desire for the Sensational," 119–121.
14. Sweeney, "The Desire for the Sensational," 119–120.
15. "Climax of Folly," *Washington Post,* May 2, 1894.
16. "Coxey Silenced by Police," *New York Times,* May 2, 1894.

17. "Coxey Silenced by Police," *New York Times,* May 2, 1894. For African Americans, this period of history has been called one of the most difficult periods in a long history of difficult periods. It was marked by rampant racism, discrimination, and mob violence. For more, see Kenneth R. Janken, "Rayford Logan: the Golden Years," *Negro History Bulletin* 61(1998); Tuskegee University Archives Online Repository, *Lyching, Whites & Negroes, 1882–1968*, 2010; and chapter 8.
18. Sweeney, "The Desire for the Sensational," 122.
19. "On the Capitol Steps," *Washington Post,* May 1, 1894.
20. Barber, *Marching on Washington*, 12.
21. Barber, *Marching on Washington*, 44.
22. Barber, *Marching on Washington*, 107.
23. Barber, *Marching on Washington*, 220.
24. "The Civil Rights Movement and Television," The Museum of Broadcast Communications, accessed December 7, 2013, http://www.museum.tv/eotvsection.php?entrycode=civilrights.
25. J. Fred MacDonald, *Blacks and White TV: Afro-Americans in Television since 1948* (Chicago, IL: Nelson-Hall Publishers, 1983), 95.
26. Juan Williams, *Eyes on the Prize: America's Civil Rights Years, 1954–1965* (New York: Penguin Books, 1988), 197–198.
27. Mark S. Greek, *Washington, D.C. Protests: Scenes from Home Rule to the Civil Rights Movement* (Charleston, S.C.: The History Press, 2009), 100–101; "Civil Rights March," NBC News Archives, August 28, 1963, accessed December 7, 2013, http://www.nbcuniversalarchives.com/nbcuni/clip/5112485941_s01.do.
28. Williams, *Eyes on the Prize*, 201.
29. Taylor Branch, *Parting the Waters: America in the King Years, 1954–1963* (New York: Simon and Shuster, 1989), 874, 879.
30. Stephanie Greco Larson, *Media & Minorities: The Politics of Race in News and Entertainment* (Lanham, MD: Rowman & Littlefield Publishers, 2006), 171.
31. Branch, *Parting the Waters,* 887.
32. Larson, *Media & Minorities,* 160.
33. Louisa Edgerly, Amoshaun Toft, and Mary Lynn Veden, "Social Movements, Political Goals, and the May 1 Marches: Communicating Protest in Polysemous Media Environments," *International Journal of Press/Politics* 16 (2011), 314–334.
34. Edgerly, Toft, and Veden, "Social Movements," 315.
35. Barber, *Marching on Washington,* 221.

Chapter 6

1. The Washington region has a relatively high percentage of foreign-born residents compared to the country at large, but the proportion who live in the actual city of Washington is not comparatively that big. "DC Mythbusting: International City," January 19, 2010, accessed October 2, 2012, http://www.welovedc.com/2010/01/19/dc-mythbusting-international-city/.

2. For this and other reasons, *Bloomberg Businessweek* ranked the city as the eleventh most global city in the world; among American cities, only New York, Los Angeles, and Chicago were ranked higher. Matt Mabe, "The World's Most Global Cities," *Businessweek,* October 29, 2008, accessed February 2, 2013, http://images.businessweek.com/ss/08/10/1028_global_cities/3.htm.
3. Donald E. Jackson, "L'Enfant's Washington: An Architect's View," *Washington History* 50 (1978): 416; Hope Ridings Miller, *Embassy Row: The Life & Times of Diplomatic Washington* (New York: Holt, Rinehart and Winston, 1969), 117. See also chapters 1 and 2.
4. Carl Abbott, *Political Terrain: Washington, D.C., from Tidewater Town to Global Metropolis* (Chapel Hill: The University of North Carolina Press, 1999), 51, 53; Miller, *Embassy Row,* 1–3; Marilyn K. Parr, "Chronicle of a British Diplomat: The First Year in the 'Washington Wilderness,'" *Washington History* 12:1 (2000), 78–89; Charles O. Paullin, "Early British Diplomats in Washington," *Records of the Columbia Historical Society* 44/45 (1942): 244; quotes from Carol M. Highsmith and Ted Landphair, *Embassies of Washington* (Washington, D.C.: Preservation Press, 1992), 13–14 and Frances Trollope, "The Domestic Manners of the Americans," excerpted from Patrick Allen, ed., *Literary Washington, D.C.* (San Antonio: Trinity University Press, 2012), 2.
5. Adam Badeau, "Gen. Badeau's Letter," *New York Times,* May 15, 1887; Carpenter, *Carp's Washington*, 70–71; Ernest R. May, *Imperial Democracy: The Emergence of America as a Great Power* (New York: Harcourt, Brace & World, 1961), 3–10; Frank Carpenter, *Carp's Washington,* (New York: McGraw-Hill, 1960), 69; Monica Hesse, "Dinner at America's Table," *Washington Post,* October 13, 2011; May, *Imperial Democracy*, 3.
6. May, *Imperial Democracy*, 5–10; "Made an Ambassador," *New York Times,* March 25, 1893; "Eustis Warmly Received," *New York Times,* May 7, 1893; "President Leaves Today," *Washington Post,* August 11, 1893; "Delighted with New-York," *New York Times,* August 25, 1893; Charles F. M. Browne, *A Short History of the British Embassy at Washington, D.C., U.S.A.* (Washington, D.C.: Gibson Bros, 1930), 17–19.
7. Bob Reinalda, *Routledge History of International Organizations: From 1815 to the Present Day* (New York: Routledge, 2009), 82. The institute, known as the Carnegie Endowment for International Peace, is headquartered just off of Dupont Circle.
8. Other early conferences included the International Labor Commission Conference in 1919 and the Washington Naval Conference in 1921–1922. Reinalda, *Routledge History of International Organizations*, 63–64, 208, 227; quote from Abbott, *Political Terrain*, 142.
9. Abbott, *Political Terrain*, 143–144.
10. Abbott, *Political Terrain*, 144–146; Hesse, "Dinner at America's Table"; Miller, *Embassy Row*, 58, 70; Larry Van Dyne, "Foreign Affairs: DC's Best Embassies," *Washingtonian Magazine*, February 1, 2008.
11. Abbott, *Political Terrain*, 143, 147–156, quote p. 156. For more on the economics of Washington, see chapter 9.

12. Van Dyne, "Foreign Affairs: DC's Best Embassies."
13. Hans Morgenthau, *Politics among Nations*, fourth edition (New York: Knopf, 1967), ch. 6.
14. Previously, it was Sixteenth Street that was known as Embassy Row; the street still features a number of embassies and ambassadorial residences (Highsmith and Landphair, *Embassies of Washington*, 14–15). England had previously helped make popular the Dupont Circle neighborhood when it opened a legation (later closed) on Connecticut and N Streets in 1872. England's was the first headquarters for a foreign diplomat built in the United States by a foreign country (Browne, *A Short History of the British Embassy*, 9).
15. Embassy of Finland website, http://www.finland.org/public/default.aspx?nodeid=35836, accessed February 1, 2013. The Canadian embassy's website, for instance, proudly proclaims that its building is "strategically situated on the processional route" between Congress and the White House, making it "one of the most prominent diplomatic presences in Washington." Embassy of Canada website, http://www.canadainternational.gc.ca/washington/offices-bureaux/about_apropos.aspx?lang=eng, accessed April 29, 2013.
16. Miller, *Embassy Row*, 12, 82; Robert Sharoff, "At the World Bank, Architecture as Diplomacy," *New York Times*, March 9, 1997.
17. Quoted in Emily Wax, "Outpost Betting a Nation," *Washington Post*, December 27, 2011.
18. As of 2011. Stephen S. Fuller, "The Economic and Fiscal Impact of Foreign Missions on the Nation's Capital," National Capital Planning Commission, June 6, 2002, p. 20, http://www.ncpc.gov/DocumentDepot/Publications/ForeignMissions/Foreign_Missions_Impact.pdf, accessed January 8, 2013.
19. International Club of D.C. website, http://www.internationalclubdc.com/Login.aspx, accessed February 2, 2013; J. Freedom du Lac, "A Chord of Jazz History to Echo at Turkish Embassy," *Washington Post,* February 4, 2011; Van Dyne, "Foreign Affairs: DC's Best Embassies." Foreign officials from nonsegregated countries found it difficult at times to abide by Washington's racial codes. In 1869, for instance, one diplomat had to leave the city after it was discovered he was associating with a local black woman. James H. Whyte, *The Uncivil War: Washington during the Reconstruction, 1865–1878* (New York: Twayne Publishers, 1958), 248. The city's segregation also created problems after World War II when embassy officials from Asia and Africa were denied hotel or restaurant service, particularly in Virginia and Maryland. In some cases, pressure from the State Department after such instances in the early 1960s led to segregation policies at "places of public accommodation" being dropped altogether. Miller, *Embassy Row*, 127–128.
20. Miller, *Embassy Row*, 5, quote p. 8; Morgenthau, *Politics among Nations*, 72–73; Highsmith and Landphair, *Embassies of Washington,* 20; Frederic Van de Water, "The Society of the Nation's Capital," in *Katharine Graham's Washington*, edited by Katharine Graham (New York: Vintage Books, 2002), 119.
21. Catherine Allgor, *A Perfect Union: Dolley Madison and the Creation of the American Nation* (New York: Henry Holt & Co., 2006), 83–101; John Newhouse,

"Diplomacy Inc.," *Foreign Affairs*, May/June 2009. Such lobbying must be done with care, however, since a perception of excessive meddling in domestic politics may offend the host country. Such was the case with Russia's minister in 1871, described above. In another example, England's emissary in 1888, Sir Lionel Sackville-West, was tricked into writing a letter that seemed to endorse President Grover Cleveland for reelection—a move that could have convinced Americans voters to vote for Cleveland's opponent, Benjamin Harrison. An irritated Cleveland ordered Sackville-West's dismissal, and the minister was recalled. May, *Imperial Democracy*, 3; "Good-Bye to Sackville," *New York Times*, October 31, 1888; "Lord Sackville's Circular," *Washington Post,* October 10, 1895. On the early efforts of foreign representatives in Washington to influence domestic politics, see James Sterling Young, *The Washington Community, 1800–1828* (New York: Columbia University Press, 1966), 219.

22. Fuller, "The Economic and Fiscal Impact of Foreign Missions"; Emily Wax, "Washington Can Be a Frontline for International Combatants," *Washington Post,* January 11, 2012.
23. Joseph T. McCann, "Spillover Effect: The Assassination of Orlando Letelier," in *Terrorism on American Soil: A Concise History of Plots and Perpetrators from the Famous to the Forgotten*, edited by Joseph T. McCann (Boulder, CO: Sentient Publications, 2006); Miller, *Embassy Row*, 133; Wax, "Washington Can Be a Frontline."
24. See, e.g., Carpenter, *Carp's Washington*, 78–79; Miller, *Embassy Row*, ch. 6; and Drew Pearson and Robert S. Allen, "The Capital Underworld," excerpted from Graham, *Personal History*, 207.
25. Miller, *Embassy Row*, 108; Mark Segraves, "Diplomats in D.C. Owe Hundreds of Thousands in Parking Tickets," *WTOP*, September 21, 2011, accessed October 2, 2012, http://www.wtop.com/41/2555713/Diplomats-in-DC-owe-hundreds-of-thousands-in-parking-tickets.
26. The official was stripped of his diplomatic immunity and given a lengthy prison sentence. Joshua E. Keating, "Can You Get Away with Any Crime If You Have Diplomatic Immunity?" *Foreign Policy*, February 15, 2011, accessed October 2, 2012, http://www.foreignpolicy.com/articles/2011/02/15/can_you_get_away_with_any_crime_if_you_have_diplomatic_immunity; "What's the Story on Diplomatic Immunity?" *The Straight Dope,* November 1, 2005, http://www.straightdope.com/columns/read/2228/whats-the-story-on-diplomatic-immunity, accessed February 2, 2013.
27. Tara Young, "Diplomat Flees U.S. to Avoid Sex Charges," *Washington Post,* March 5, 2005.
28. Jack Eisen and LaBarbara Bowman, "House Joins Senate to Kill Chancery Bill," *Washington Post,* December 21, 1979; Highsmith and Landphair, *Embassies of Washington*, 16; Miller, *Embassy Row*, 116, 118, 121; Van Dyne, "Foreign Affairs: DC's Best Embassies."
29. For more on Pack, see Mary S. Lovell, *Cast No Shadow: The Life of the American Spy Who Changed the Course of World War II* (New York: Pantheon Books, 1992).

Chapter 7

1. Thanks to Tara Hamilton for this important insight. Washington's independence from a county resembles that of other cities that have consolidated with counties, such as Nashville and Louisville.
2. U.S. Constitution, Art. I sec. 8. Federal law also allows the national government to exercise exclusive jurisdiction over certain kinds of property, such as military facilities; see 18 U.S.C. § 7, section 7(3). Most cities in the United States have only limited independence: under the so-called Dillon's Rule, their authority derives from the state and can, in theory, be taken away by law or constitutional amendment. Edward C. Banfield and James Q. Wilson, *City Politics* (Cambridge, MA: Harvard University Press and the M.I.T. Press, 1963), 63–67; Dennis R. Judd and Todd Swanstrom, *City Politics: The Political Economy of Urban America*, 5th ed. (New York: Pearson/Longman, 2006), 38–39; Paul E. Peterson, *City Limits* (Chicago: The University of Chicago Press, 1981).
3. This was the finding of a 2007 study that compared Washington to several other capital cities, including those of Australia, Belgium, and Germany. Hal Wolman et al., "Capital Cities and Their National Governments: Washington, D.C. in Comparative Perspective," GWIPP Working Paper 30, June 11, 2007. Nigeria probably comes closest to the United States in terms of limiting the independence of its capital. J. Isawa Elaigwu, "Abuja, Nigeria," in *Finance and Governance of Capital Cities in Federal Systems*, ed. Enid Slack and Rupak Chattopadhyay (Montreal: McGill-Queen's University Press, 2009), 209–211.
4. Libby Copeland, "Shadow Delegation Toils in Obscurity for D.C.'s Day in the Sun," *Washington Post*, January 16, 2007. At present the district's delegate to the House may participate in votes within her committees but cannot cast ballots on the floor of the chamber. The use of shadow representatives copies a tactic used by Tennessee to lobby for full statehood in the 1790s. Edward M. Meyers, *Public Opinion and the Political Future of the Nation's Capital* (Washington, D.C.: Georgetown University Press, 1996), 39.
5. Charles Wesley Harris, *Congress and the Governance of the Nation's Capital: The Conflict of Federal and Local Interests* (Washington, D.C.: Georgetown University Press, 1995), 60; Constance McLaughlin Green, *Washington, Vol. 1: Village and Capital, 1800–1878* (Princeton, NJ: Princeton University Press, 1962), 66; Enid Slack and Rupak Chattopadhyay, "Finance and Governance of Capital Cities in Federal Systems," *Washington, D.C. Economic Partnership Forum*, September 16, 2010, accessed January 15, 2012, http://www.youtube.com/watch?v=XG-fY3HNuFU&feature=related., 9:00 to 11:27; see also Wolman et al., "Capital Cities."
6. Alexander Hamilton, James Madison, and John Jay, *The Federalist Papers*, New York: Signet Classics (2003 [1788]), no. 43; see also Green, *Washington, Vol. 1*, 11. There were some who worried, both during the drafting of the Constitution and debates over its ratification, that Congress's power over the district was too great (Kenneth R. Bowling, *The Creation of Washington, D.C.: The Idea and Location of the American Capital* (Fairfax, VA: George Mason University Press, 1991), 76–79, 83–85).

7. This guarantee of suffrage rights was added by James Madison. Interestingly, by doing so he forced a delay in approving the bill to place the capital city—a delay that allowed Madison and other Southerners to successfully lobby for the city to be placed on the Potomac River, rather than in Germantown, Pennsylvania. Bowling, *The Creation of Washington, D.C.*, 158–160, 190, 212–213.
8. Viet D. Dinh and Adam H. Charnes, "The Authority of Congress to Enact Legislation to Provide the District of Columbia with Voting Representation in the House of Representatives," testimony submitted to the Committee on Government Reform, U.S. House of Representatives, 2004, accessed November 22, 2011, http://www.dcvote.org/trellis/research/finaldinhopinion.cfm. The three cities originally within the District of Columbia—Georgetown, Alexandria, and Washington—each had their own local government. The portion of the district that contained Alexandria became part of Virginia in the 1840s, and the 1871 act that created a new commissioner form of government merged Georgetown, Washington, and D.C. into a single municipal unit, though they were not fully consolidated until 1895. See Robert Harrison, *Washington during Civil War and Reconstruction: Race and Radicalism* (New York: Cambridge University Press, 2011), ch. 8.
9. The Board of Trade was formed in 1889; a second influential business group, the Federal City Council, was formed in the 1950s. Both groups exercised influence by forming and exploiting personal connections with members of the House District Committee. Martha Derthick, *City Politics in Washington, D.C.*, Joint Center for Urban Studies of the Massachusetts Institute of Technology and Harvard University (Cambridge: Harvard University Press, 1962), 57–61, 87–88; Michael K. Fauntroy, *Home Rule or House Rule? Congress and the Erosion of Local Governance in the District of Columbia* (Dallas, TX: University Press of America, 2003), 46–50; Alan Lessoff, *The Nation and Its City: Politics, "Corruption," and Progress in Washington, D.C., 1861–1902* (Baltimore, MD: Johns Hopkins University Press, 1994), ch. 5, 209–225; see also chapter 9 and Constance McLaughlin Green, *Washington, Vol. 2: Capital City, 1879–1950* (Princeton, NJ: Princeton University Press, 1963), 29–34, 37. Commission government was employed by other cities during a recession in the 1870s, and again at the turn of the century when technocratic, business-like principles of governance were popular. Judd and Swanstrom, *City Politics*, 93–97; Lessoff, *The Nation and Its City*, 204; Kate Masur, 2010, *An Example for All the Land: Emancipation and the Struggle over Equality in Washington, D.C.*, Chapel Hill: University of North Carolina Press, 250. Lessoff (1994) argues that D.C.'s commission government was more akin in form to the latter than the former, while the political power of the Board of Trade resembled the influence of Southern business alliances (pp. 123, 209).
10. The push for district representation in the federal government began in the 1880s by local reporter and future newspaper editor Theodore Noyes, and organized efforts to restore home rule began as early as the 1910s. Derthick, *City Politics in Washington, D.C.*, 73; Mark S. Greek, *Washington, D.C. Protests: Scenes from Home Rule to the Civil Rights Movement* (Charleston, S.C.: The History Press, 2009); Green, *Washington, Vol. 2*, 25–26.

11. Fauntroy, *Home Rule or House Rule?*; Harris, *Congress and the Governance of the Nation's Capital*; Phillip G. Schrag, "The Future of District of Columbia Home Rule," *Catholic University Law Review* 39:2 (1990), 329–330, 342. Congress has rarely been successful at explicitly overturning city laws, preferring instead its power over the city's budget to accomplish the same goal (e.g. Harris, *Congress and the Governance of the Nation's Capital,* chs. 4 and 5).
12. Schrag, "The Future of District of Columbia Home Rule," 342; Neil Spitzer, "A Secret City," *Wilson Quarterly* 13:1 (New Year's 1989), 103.
13. Harrison, *Washington during Civil War*, 298–301.
14. Harrison, *Washington during Civil War*, 277–278, 294–295; Lessoff, *The Nation and Its City*, 52–53, 119; Kate Masur, *An Example for All the Land: Emancipation and the Struggle over Equality in Washington, D.C.* (Chapel Hill: University of North Carolina Press, 2010), 251–253, quote p. 251. When congressional Democrats first considered a bill in 1878 that gave the city a weak city council, they tried to impose a poll tax and property requirement for voting, guaranteeing that all but a few of the city's black residents would be unable to vote (Harrison, *Washington during Civil War*, 298–301; see also *Congressional Record*, March 20, 1878, p. 1922, and May 7, 1878, p. 3247). The partial repeal of local government for the city in 1871 was seen by Frederick Douglass and other prominent African American Washingtonians as a step toward denying blacks the right to vote, as well as a sign of hope for local whites "who had carped about Washington's 'negro government'" (Green, *Washington, Vol. 1*, 337).
15. Fauntroy, *Home Rule or House Rule?,* 52; Jay Franklin, "Main Street-on-Potomac," in *Katharine Graham's Washington*, edited by Katharine Graham (New York: Vintage Books, 2002), 23; Green, *Washington, Vol. 1*, 361; Lessoff, *The Nation and Its City*, 103; Mondell, 1938. Even today, lawmakers who are what political scientist Michael Fauntroy calls "racial conservatives" may gain plaudits from like-minded constituents for attacking the D.C. government and its African American leaders (Fauntroy, *Home Rule or House Rule?*, 90).
16. Congressional resistance to D.C. suffrage was especially strong in the House's District Committee, though some Southerners on that committee were considered friendlier to the city than other Southern lawmakers. Derthick, *City Politics in Washington, D.C.*, 53–54; Steven J. Diner, "The City under the Hill," *Washington History* 8:1 (Spring/Summer 1996), p. 58. Not coincidentally, all Southern states but Tennessee refused to ratify the Twenty-Third Amendment to the Constitution in 1961, giving the district the right to vote in presidential elections.
17. Fauntroy, *Home Rule or House Rule?*, 51. The Democratic leanings of urban and black voters first emerged in the 1930s. Judd and Swanstrom, *City Politics,* 110, 117, 122. For more on race in Washington, see chapter 8.
18. Green, *Washington, Vol. 1*, 357–360; Lessoff, *The Nation and Its City*, 76; Masur, *An Example for All the Land*, 248–250; James H. Whyte, *The Uncivil War: Washington during the Reconstruction, 1865–1878* (New York: Twayne Publishers, 1958), 80–83, 91. This helps explain why a Republican-led Congress agreed to end local elections in a G.O.P.-majority city. In fact, according to the congressional

committee that proposed ending local government, the move was intended to be a short-term fix because there was not enough time before adjournment for Congress to enact a new and more viable system of local government (Whyte, *The Uncivil War*, 267; see also pp. 226 and 231).

19. Quote from Green, *Washington, Vol. 2*, 5; Whyte, *The Uncivil War*, 212–213, 227; Fauntroy, *Home Rule or House Rule?*, 13 and ch. 6; Meyers, *Public Opinion*, 29; see also pp. 69–70.
20. Greek, *Washington, D.C. Protests*, 14; see also Lessoff, *The Nation and Its City*, 115, 199–200. In 1979, for instance, Congress overturned a local law that limited where foreign chanceries could be built in the city, which could have hampered the country's relations with foreign countries (Harris, *Congress and the Governance of the Nation's Capital*, 101–111). See Chapter 8.
21. See also chapter 9.
22. Kenneth R. Bowling, "From 'Federal Town' to 'National Capital': Ulysses S. Grant and the Reconstruction of Washington, D.C.," *Washington History* 14:1 (Spring/Summer 2002), 14; Fauntroy, *Home Rule or House Rule?*, 26–27; Green, *Washington, Vol. 2*, 24, 56–57, 180; Kent Jenkins, Jr., "D.C.'s Clout Dwindling on the Hill, Lawmakers Warn," *Washington Post*, July 5, 1993; Howie Kurtz and Michael Isikoff, "Congress Still Rules the Roost in District," *Washington Post*, October 25, 1981; Meyers, *Public Opinion*, 26. See also Harris, *Congress and the Governance of the Nation's Capital* for other examples. Representatives and senators from Maryland and Virginia frequently sat on D.C. oversight committees in the late nineteenth century and used their positions to help their states (Lessoff, *The Nation and Its City*, 158).
23. Mona E. Dingle, "*Gemeinschaft und Gemutlichkeit*: German American Community and Culture, 1850–1920," in *Urban Odyssey: A Multicultural History of Washington, D.C.*, ed. Francine Curro Cary (Washington, D.C.: Smithsonian Institution Press, 130); Constance McLaughlin Green, *The Secret City: A History of Race Relations in the Nation's Capital* (Princeton: Princeton University Press, 1967), 59–60; see also chapter 8. Ironically, Truman's support for desegregating Washington led some city whites to lodge the same complaint that blacks had once made: that the city lacked local governance and was an easy target for national politicians. See, e.g., Bell Clement, "Pushback: The White Community's Dissent from 'Bolling,'" *Washington History* 16:2 (Fall/Winter 2004): 93.
24. Kevin J. McMahon, *Nixon's Court: His Challenge to Judicial Liberalism and Its Political Consequences* (Chicago: University of Chicago Press, 2011), 189–190; Thomas G. Smith, *Showdown: JFK and the Integration of the Washington Redskins* (Boston: Beacon Press, 2011). When Congress follows a more hands-off approach toward the city, it is often because it agrees with existing city policies. In the 1980s, for example, congressional Democrats who supported home rule also "wanted the District government to succeed . . . to be a showcase of their liberal programs." Harry S. Jaffe and Tom Sherwood, *Dream City: Race, Power, and the Decline of Washington, D.C.* (New York: Simon & Schuster, 1994), 133.

25. Derthick, *City Politics in Washington, D.C.*, 57; see also Harrison, *Washington during Civil War*, 109–110. Constituent pressure on Congress to change the district's laws goes back to the early years of the American Republic. Before the Civil War, for instance, Congress received countless petitions from antislavery advocates urging the abolition of slavery in the city. Josephine F. Pacheco, *The Pearl: A Failed Slave Escape on the Potomac* (Chapel Hill: University of North Carolina Press, 2005), 9–10. More examples of such external lobbying are mentioned in Green, *Washington, Vol. 2*, 39, 42.
26. Harris, *Congress and the Governance of the Nation's Capital*, 59, 73, 112–118; Eleanor Holmes Norton, "Home Rule Doesn't Come with an Asterisk," *Washington Post*, April 24, 2011; Schrag, "The Future of District of Columbia Home Rule," 313; see also Kurtz and Isikoff, "Congress Still Rules the Roost." Similarly, in the 1860s, city Republicans asked Congress to change the city's charter to give them more power in local government, and the G.O.P. altered election laws in 1867 to improve their party's chances in city elections (Harrison, *Washington during Civil War*, 164–165, 168). Given the prevailing view in Washington politics that congressional interference is unfair, this belief among Washington denizens that Congress is needed to protect the city from bad local decisions "tends to be whispered [in Washington] but never publicly voiced" (Meyers, *Public Opinion*, 71).
27. Tim Craig, "Gridlock? On D.C. Issues, Gray and Issa Defy Spirit of the Times," *Washington Post*, April 22, 2012; Schrag, "The Future of District of Columbia Home Rule," 313n16, 316; quoted from Spitzer, "A Secret City," 103; see also Harris, *Congress and the Governance of the Nation's Capital*, 160.
28. Howard Gillette, Jr., *Between Justice and Beauty: Race, Planning, and the Failure of Urban Policy in Washington, D.C.* (Baltimore: Johns Hopkins University Press, 1995), 207; Jaffe and Sherwood, *Dream City*, 319. Not all city officials agree on the best form that independence should take; some have supported statehood for the district, others a lesser degree of autonomy (see, e.g., Clifford D. May, "Washington Talk: Home Rule," *New York Times*, January 11, 1989).
29. Daniel Newhauser, "Gray Arrest Highlights D.C. Battle," *Roll Call*, April 12, 2011. Gray was following in the footsteps of another mayor, Sharon Pratt Kelly, who did the same thing in 1993.
30. There are dangers with this strategy. In 1993, for instance, the city's budget nearly failed in the House, partly because then-Mayor Sharon Pratt Kelly had "used [members of Congress] as a scapegoat for a host of city problems" (Jenkins, "D.C.'s Clout Dwindling").
31. Kurtz and Isikoff, "Congress Still Rules the Roost"; Jaffe and Sherwood, *Dream City*; Eric Pianin and Saundra Torry, "Barry Seeks to Quell Anger of D.C. Officials," *Washington Post*, October 4, 1988; Steve Twomey, "Blacks Make Barry the People's Chance," *Washington Post*, September 15, 1994.
32. See, e.g., Pianin and Torry, "Barry Seeks to Quell Anger." Barry did not limit his alliances to Democrats; he even tried, albeit briefly, to forge an alliance with then-Speaker Newt Gingrich in 1995. Jonetta Rose Barras, 1998, *The Last of the Black*

Emperors: The Hollow Comeback of Marion Barry in the New Age of Black Leaders (Baltimore: Bancroft Press, 1998), 186.

33. See, for example, Barras, *The Last of the Black Emperors,* 249–250; quote from Spitzer, "A Secret City," 107.
34. Derthick, *City Politics in Washington, D.C.*, 95–100; Harris, *Congress and the Governance of the Nation's Capital,* 6; Jaffe and Sherwood, *Dream City*, 27–28, 62.
35. Derthick, *City Politics in Washington, D.C.*, 138–139; Steven J. Diner, "From Jim Crow to Home Rule," *The Washington Quarterly* 13:1 (New Year's 1989), 96, 98; Diner, "The City under the Hill," 58; Fauntroy, *Home Rule or House Rule?*, 41–43; Jaffe and Sherwood, *Dream City*, 34, 41, 45–47; Peter Milius, "Why Coates? Image . . . Unity," *Washington Post*, January 29, 1969; "Negro May Head D.C. School Board," *New York Times,* November 28, 1968.
36. Katherine Tate, *From Protest to Politics: The New Black Voters in American Politics* (New York: Russell Sage Foundation, 1993); quote from Michael C. Dawson, *Behind the Mule: Race and Class in African-American Politics* (Princeton, NJ: Princeton University Press, 1994), 5. That strong group consciousness naturally follows from centuries in which "race was the decisive factor in determining the opportunities and life chances available to virtually all African Americans" (Dawson, *Behind the Mule*, 10). But certain factors, such as individual socioeconomic status and the characteristics of one's neighborhood, can modify how strongly individual blacks hold a racial identity. See, e.g., Claudine Gay, "Putting Race in Context: Identifying the Environmental Determinants of Black Racial Attitudes," *American Political Science Review* 98:4 (2004); Donald Kinder and Nicholas Winter, "Exploring the Racial Divide: Blacks, Whites, and Opinion on National Policy," *American Journal of Political Science* 45 (April 2011).
37. J. Phillip Thompson III, *Double Trouble: Black Mayors, Black Communities, and the Call for a Deep Democracy* (New York: Oxford University Press, 2006), 32. For instance, after his narrow win for mayor in 1978, Barry focused on "shoring up his African-American power base" and often made "appeals to racial solidarity," not an uncommon technique for black mayors at the time. Howard Gillette, Jr. "Protest and Power in Washington, D.C.: The Troubled Legacy of Marion Barry," in *African-American Mayors: Race, Politics, and the American City*, edited by David R. Colburn and Jeffrey S. Adler (Urbana, Ill: University of Illinois Press, 2001), 204. For other examples, see Jaffe and Sherwood, *Dream City,* 196 and Spitzer, "A Secret City," 107.
38. Derthick, *City Politics in Washington, D.C.*, 68, 130; Sam Fulwood III, "Why Blacks Support Mayor Barry," *Los Angeles Times,* August 11, 1990; Jaffe and Sherwood, *Dream City*, 30, 132–133, 181, 190; Eric Pianin and Courtland Milloy, "Does the White Return to D.C. Mean 'The Plan' Is Coming True?" *Washington Post,* October 6, 1985; quote from Barras, *The Last of the Black Emperors,* 79.
39. Jaffe and Sherwood, *Dream City*, 140–142, 189; Steven Taylor, "Political Culture and African Americans' Forgiveness of Elected Officials," *Polity* 37:4 (October 2005); Thompson, *Double Trouble*, 62–67. See also Dennis E. Gale, *Washington,*

D.C.: Inner-City Revitalization and Minority Suburbanization (Philadelphia: Temple University Press, 1987), 176–177 and Michael Janofsky, "The 1994 Campaign: The Comeback Man in the News,'" *New York Times*, September 14, 1994. Barry's tight control of patronage and other benefits meant city council members were "dependent on [Barry] for constituent services," allowing Barry to garner their support for his initiatives (Jaffe and Sherwood, *Dream City*, 180). It was not until Barry was able to win election without the city's white vote, starting in 1982, that he became known as a more race-oriented "Boss Barry" (Jaffe and Sherwood. *Dream City*, 143). For a concise and even-handed account of Barry's long rule as mayor, see Gillette, Jr., "Protest and Power in Washington, D.C."

40. Nikita Stewart and Paul Schwartzman, "How Adrian Fenty Lost His Reelection Bid for D.C. Mayor," *Washington Post*, September 16, 2010.
41. Jaffe and Sherwood, *Dream City*, 47–48, 98, 123, 140; quote by former D.C. official, March 2012. Barras (1998) offers a more critical look at Barry's political skills. On Barry's campaign abilities, see Jaffe and Sherwood, *Dream City*, and Rene Sanchez, "Barry Comes Roaring Back in D.C.," *Washington Post*, September 14, 1994.
42. Economic divisions are hardly unique to Washington. See, for example, Peter Dreier, John H. Mollenkopf, and Todd Swanstrom, *Place Matters: Metropolitics for the Twenty-First Century* (Lawrence, KS: University of Kansas Press, 2005), xv; and Banfield and Wilson, *City Politics*, 35–37. Within the black community, "the strongest supporters of black mayors tend to come from the black middle class, not the poor," and upper- and middle-class African Americans have often had different views on urban issues than poorer blacks (Banfield and Wilson, *City Politics*, 298; quote from Thompson, *Double Trouble*, 48). In 2010, Ward 8 was 94 percent black, and in 1999 had an average family income of $47,000; Ward 3 was 5.6 percent black in 2010 and in 1999 had an average family income of $245,000. Urban Institute, "Neighborhood Profiles," accessed January 9, 2013, http://www.neighborhoodinfodc.org/ wards/wards.html.2012. The boundaries of the wards have changed somewhat, albeit not dramatically, since the 1970s.
43. This occurred, for instance, in Sharon Pratt Kelly's 1990 primary election (Richard Morin and Michael Abramowitz, "Dixon Victory Points to Economic Rift between Voters," *Washington Post*, September 16, 1990).
44. Thompson, *Double Trouble*, 4, 6–8, 13, quote p. 6. For more on minority elected officials in various city governments, see Judd and Swanstrom, *City Politics*, 399–409.
45. Judd and Swanstrom, *City Politics*, 344–346.
46. John Guinther, *Direction of Cities* (New York: Penguin Books, 1996), 169; Judd and Swanstrom, *City Politics*, 331, 338–343, quote p. 343; Peterson, *City Limits*, 29, 32, 64, quotes pp. 4, 22.
47. By one estimate, more than 56 percent of district land area is tax exempt (Fauntroy, *Home Rule or House Rule?*, 151).
48. Jon Bouker et al., *Building the Best Capital City in the World* (Brookings Institution, Washington, D.C., 2008), 25; Emma Dumain, "Commuter Tax Is Panned," *Roll Call*, July 25, 2012. This has long been a problem plaguing Washington; see,

for example, Harrison, *Washington during Civil War*, 23–25. John Guinther calls this the "mainstreaming" of elected officials, which affected new black political leaders but can impact incumbents regardless of race (Guinther, *Direction of Cities*, 126; see also Thompson, *Double Trouble*, 41). For more on the city's unique fiscal challenges, see Natwar M. Gandhi et al., "Washington, District of Columbia, United States of America," in *Finance and Governance of Capital Cities in Federal Systems*, ed. Enid Slack and Rupak Chattopadhyay (Montreal: McGill-Queen's University Press, 2009).

49. Kenneth T. Jackson, *Crabgrass Frontier: The Suburbanization of the United States* (New York: Oxford University Press, 1985), 138–140, 148–150.
50. Banfield and Wilson, *City Politics*, 15; Gillette, Jr., *Between Justice and Beauty*, 154; Judd and Swanstrom, *City Politics*, 236, 244; quote from Guinther, *Direction of Cities*, 157.
51. Fauntroy, *Home Rule or House Rule?* 26, 54–57, 116–118; Whyte, *The Uncivil War*, 160. The inability to tax federal property was a source of frustration from the city's early years (e.g., Green, *Washington, Vol. 1*, 39). Other cities have been able to impose commuter taxes, though not many (Judd and Swanstrom, *City Politics*, 340).
52. Natwar M. Gandhi, "Tax Rates and Tax Burdens in the District of Columbia—A Nationwide Comparison, 2010," Office of the Chief Financial Officer, Government of the District of Columbia, 2010, http://www.cfo.dc.gov/cfo/frames.asp?doc=/cfo/lib/ cfo/10study.pdf. Shared authority over the city's budget also means there is no clear governing responsibility for how the city manages itself, allowing city and congressional leaders to point fingers at each other "when things go wrong." Charles Wesley Harris, "In Whose Interest? Congressional Funding for Washington in the Home-Rule Era," *Washington History* (Spring/Summer 1996), 66.
53. Meyers, *Public Opinion*, 27–29. Gillette, 1995, provides a fine historical review of Congress's changing financial commitment to the district; see also Green, *Washington, Vol. 1* and *Washington, Vol. 2* and Harrison, *Washington during Civil War*.
54. Gillette, Jr., *Between Justice and Beauty*, 64–66, 70; Harrison, *Washington during Civil War*, 302–303; Lessoff, *The Nation and Its City*, 103. This desire for congressional aid helped maintain the popularity of congressional governance through the rest of the nineteenth century (Lessoff, *The Nation and Its City*, 203). A majority of African Americans in the city opposed the 1995 control board, whereas D.C. whites supported it by an even wider margin (Fauntroy, *Home Rule or House Rule?* 173).
55. See, e.g., R. Douglas Arnold and Nicholas Carnes, "Holding Mayors Accountable: New York's Executives from Koch to Bloomberg," *American Journal of Political Science* 56:4 (2012). The Washington economy is discussed further in chapter 9.
56. Fauntroy, *Home Rule or House Rule?* 140–143; Jaffe and Sherwood, *Dream City*, 187; Spitzer, "A Secret City," 110–112. By the mid-1990s, a majority of city residents saw their city as inefficient and corrupt (Meyers, *Public Opinion*, 41–42).

For more on corruption and poor city services under the Barry administration, see Jaffe and Sherwood, *Dream City*, esp. 151–155, 186, and 209–211.

57. Robert McCartney, "A Warning to District's Political Old Guard," *Washington Post,* May 2, 2013. For more, see chapter 8.
58. Mike DeBonis and Ben Pershing, "Vote Didn't End Budget-Autonomy Debate," *Washington Post,* April 28, 2013; Fauntroy, *Home Rule or House Rule?* 2, 36–37; Meyers, *Public Opinion*, 174, quote p. 34. In 1978, Congress passed a constitutional amendment giving the district voting representation in the House and Senate but failed to win approval of enough states. In 2010, the House came close to passing a bill giving both the district and the state of Utah an additional representative, but it too failed.

Chapter 8

1. Carol Morello and Dan Keating, "Blacks' Majority Status Slips Away," *Washington Post,* March 25, 2011.
2. Natalie Hopkinson, *Go-Go Live: The Musical Life and Death of a Chocolate City* (Durham, NC: Duke University Press, 2012), xiii.
3. See, for example, Theodore W. Allen, *The Invention of the White Race: The Origin of Racial Oppression in Anglo America* (New York: Verso, 1994); Matthew Frye Jacobson, *Whiteness of a Different Color: European Immigrants and the Alchemy of Race* (Cambridge: Harvard University Press, 1999).
4. There were nearly 288,000 slaves in Virginia in 1790, 42 percent of its total population. Campbell Gibson and Kay Jung, *Historical Census Statistics on Population Totals by Race, 1790 to 1990, and by Hispanic Origin, 1970 to 1990, for the United States, Regions, Divisions, and States,* Working Paper Series No. 56 (Washington D.C.: U.S. Census, September 2002), table 61.
5. Letitia Woods Brown, *Free Negroes in the District of Columbia, 1790-1846* (New York: Oxford University Press, 1972); Jesse J. Holland, *Black Men Built the Capitol: Discovering African-American History in and around Washington, D.C.* (Guilford, CT: The Globe Pequot Press, 2007); Gibson and Jung, *Historical Census Statistics for the United States,* table 23; Constance McLaughlin Green, *Washington, Vol. 1: Village and Capital, 1800–1878* (Princeton, NJ: Princeton University Press, 1962), 144; Stanley Harrold, *Subversives: Antislavery Community in Washington, D.C., 1828–1865* (Baton Rouge: Louisiana State University Press, 2003), 15, 101; Clarence Lusane, *The Black History of the White House* (San Francisco: City Lights Books, 2011); Manuel Roig-Franzia, "Researcher Links Slaves to Castle's Sandstone," *Washington Post,* December 13, 2012. In 1806 Virginia passed a law that prohibited former slaves from living there for more than a year, compelling many to flee to D.C.
6. Brown, *Free Negroes*, 62–63, 129–135; Ernest B. Furgurson, *Freedom Rising: Washington in the Civil War* (New York: Alfred A. Knopf, 2004), 12–13; Constance McLaughlin Green, *The Secret City: A History of Race Relations in the Nation's Capital* (Princeton: Princeton University Press, 1967), 16, 18–19, 25,

32, 40–43; James Oliver Horton, "The Genesis of Washington's African American Community," in *Urban Odyssey: A Multicultural History of Washington, D.C.*, ed. by Francine Curro Cary (Washington, D.C.: Smithsonian Institution Press), 23–27, 30–32; Allan Johnston, *Surviving Freedom: The Black Community of Washington, D.C., 1860–1880* (New York: Garland Publishing, 1993), 88–89, 101; Dorothy Provine, 1973, "The Economic Position of Free Blacks in the District of Columbia, 1800–1860," *The Journal of Negro History* 58:1 (1973); James H. Whyte, *The Uncivil War: Washington During the Reconstruction, 1865–1878* (New York: Twayne Publishers, 1958), 27–28. Schooling was available for blacks, however, and the black literacy rate was high compared to blacks in Southern cities. Robert Harrison, *Washington during Civil War and Reconstruction: Race and Radicalism* (New York: Cambridge University Press, 2011), 9. Georgetown adopted its own black codes in 1832 (Green, *The Secret City*, 35). Wilson was later freed with the help of the owner of the boardinghouse where Wilson worked (Harrold, *Subversives*, 108–111).

7. Douglas E. Evelyn and Paul Dickson, *On This Spot: Pinpointing the Past in Washington, D.C.* (Washington, D.C.: Farragut Publishing Company, 1992), 27; Furgurson, *Freedom Rising*, 99–100; Harrold, *Subversives*, 7; Horton, "The Genesis of Washington's African American Community," 25–26; Lusane, *The Black History of the White House*, 106.
8. Brown, *Free Negroes*, 124–125; Harrold, *Subversives*, 5, 31, 34–35, 256, ch. 3, quote p. 30; Horton, "The Genesis of Washington's African American Community," 37–38. For more on the Underground Railroad of Washington, see Furgurson, *Freedom Rising*, 102–105.
9. Several local blacks may have also helped organize the escape, including Paul Jennings, former slave of President James Madison's. Harrold, *Subversives*, 97–98, 128; Josephine F. Pacheco, *The Pearl: A Failed Slave Escape on the Potomac* (Chapel Hill: University of North Carolina Press, 2005); Mary Kay Ricks, "Escape on the Pearl," *Washington Post,* August 12, 1998. The *Pearl* incident also frightened local slave owners into selling their slaves, as did fears of an impending ban on the slave trade in the district (Harrold, *Subversives*, 148).
10. Noel Ignatiev, *How the Irish Became White* (New York: Routledge, 1995); David R. Roediger, *The Wages of Whiteness: Race and the Making of the American Working Class* (New York: Verso, 1999), ch. 7. For more on how European immigrants became classified as "white" over time, see Jacobson, *Whiteness of a Different Color*.
11. Harrold, *Subversives*, 33. The riot was named after Beverly Snow, the black owner of a downtown restaurant who fled the violence and eventually moved to Canada. Jefferson Morley, *Snow-Storm in August: Washington City, Francis Scott Key, and the Forgotten Race Riot of 1835* (New York: Doubleday, 2012).
12. Green, *The Secret City*, 46; Harrold, *Subversives*, 5; Jeffrey A. Jenkins and Charles Stewart, III, *Fighting for the Speakership: The House and the Rise of Party Government* (Princeton: Princeton University Press, 2013), 34. Local freedmen tried to impede efforts to enforce the Fugitive Slave Act, publicly shaming slave traders

who tried to capture runaway slaves. Kate Masur, *An Example for All the Land: Emancipation and the Struggle over Equality in Washington, D.C.* (Chapel Hill: University of North Carolina Press, 2010), 28–29.

13. A. Glenn Crothers, "The 1846 Retrocession of Alexandria: Protecting Slavery and the Slave Trade in the District of Columbia," in *In the Shadow of Freedom: The Politics of Slavery in the National Capital*, ed. by Paul Finkelman and Donald R. Kennon (Athens: Ohio University Press, 2011); Harrold, *Subversives*, 44–45, 68, quote p. 44; James B. Stewart, "Christian Statesmanship, Codes of Honor, and Congressional Violence: The Antislavery Travails and Triumphs of Joshua Giddings," from *In the Shadow of Freedom*. Congress did not ban the sale of slaves to and from district residents, however, and such trade continued, while some slave traders simply relocated across the Potomac River to Alexandria. Harrold, *Subversives*, 166.
14. Green, *The Secret City*, 60; Harrison, *Washington during Civil War*, 118; Masur, *An Example for All the Land*, 26–27; quote from Furgurson, *Freedom Rising*, 220.
15. Johnston, *Surviving Freedom*, 112–113, 125, 135–136. For more on Washington during the Civil War, see Margaret Leech, *Reveille in Washington: 1860–1865* (New York: London, Harper & Bros, 1942).
16. Another 8,000 blacks lived in the district but outside of Washington City in 1870. Gibson and Jung, *Historical Census Statistics for the United States*; Campbell Gibson and Kay Jung, *Historical Census Statistics on Population Totals by Race, 1790 to 1990, and by Hispanic Origin, 1970 to 1990, for Large Cities and Other Urban Places in the United States*, Working Paper No. 76 (Washington, D.C.: U.S. Census, February 2005), table 9; Green, *The Secret City*, 61–64. A national political culture of self-reliance also restrained what government could do for them, though some assistance was provided. Johnston, *Surviving Freedom*, chs. 6–7.
17. Green, *The Secret City*, 94–96, quote p. 76; Masur, *An Example for All the Land*, 131–146, 158–161. Additional antidiscrimination bills were enacted by the D.C. government in 1872 and 1873; a national law forbidding discrimination, passed in 1875, was overturned by the Supreme Court (Green, *The Secret City*, 108–109; Whyte, *The Uncivil War*, 246–248). For more on local government's opposition to equal rights or assistance for African Americans, see Masur, *An Example for All the Land*, 54–59, 78.
18. Harrison, *Washington during Civil War*, 2011; Masur, *An Example for All the Land*, 146–148. Washington Mayor Sayles J. Bowen in particular (1868–1870) appointed blacks to city positions and created jobs for black laborers. Harrison, *Washington during Civil War;* Masur, *An Example for All the Land*.
19. Johnston, *Surviving Freedom*, 60; Masur, *An Example for All the Land*, 7, 65, 77–85, ch. 3; Jacqueline M. Moore, *Leading the Race: The Transformation of the Black Elite in the Nation's Capital, 1880–1920* (Charlottesville, VA: University Press of Virginia, 1999), 24; Whyte, *The Uncivil War*, 258–261.
20. James Borchert, *Alley Life in Washington: Family, Community, Religion, and Folklife in the City, 1850–1970* (Urbana, IL: University of Illinois Press, 1980),

237; Spencer R. Crew, "Melding the Old and the New: The Modern African American Community, 1930–1960," in *Urban Odyssey*, 213; Johnston, *Surviving Freedom*, 5–9, 12–17, 160–64; Green, *The Secret City*, 81, 127; Whyte, *The Uncivil War*, 257. Some alleyways had been home to large concentrations of African Americans before the Civil War, too (Borchert, *Alley Life in Washington*, 6).

21. Harrison, *Washington during Civil War*, 142, 272, 303–304, 307; Masur, *An Example for All the Land*, 188–194, 208–212, 259; Whyte, *The Uncivil War*, 270. See also chapter 7. African Americans did occasionally lobby the unelected commission governing the city on specific issues. Alan Lessoff, *The Nation and Its City: Politics, 'Corruption,' and Progress in Washington, D.C., 1861–1902* (Baltimore, MD: Johns Hopkins University Press, 1994), 202.
22. Blacks also attempted to organize banks and loan societies in the late nineteenth century, with mixed success. Green, *The Secret City*; Moore, *Leading the Race*, 12, 137–139; John Muller, *Frederick Douglass in Washington, D.C.: The Lion of Anacostia* (Charleston, SC: History Press, 2012).
23. Green, *The Secret City*, 24–25, 40, 150–151; Johnston, *Surviving Freedom*, 62; Moore, *Leading the Race*, 16–21, ch. 8; Moore, *Leading the Race*, 15, 21, 27; Whyte, *The Uncivil War*, 261. The high reputation of Dunbar High School led some blacks to move to Washington just so their children could attend (Crew, "Melding the Old and the New," 211).
24. Borchert, *Alley Life in Washington*, 82, 103; Johnston, *Surviving Freedom*, 107; Moore, *Leading the Race*, 2, 5–7, 17, 24, 31.
25. Mara Cherkasky, "'For Sale to Colored': Racial Change on S Street, N.W.," *Washington History* 8:2 (1996), 43; Gary Gerstle, *American Crucible: Race and Nation in the Twentieth Century* (Princeton: Princeton University Press, 2001); Green, *The Secret City*, 155–159, 165–166; Moore, *Leading the Race*, 152–153, 156.
26. Lusane, *The Black History of the White House*, 225–231; quote from Green, *The Secret City*, 126.
27. Mary Church Terrell, "What It Means to Be Colored in Capital of U.S," 1906, accessed November 20, 2012, http://www.americanrhetoric.com/speeches/marychurchterellcolored.htm.
28. "Insist upon Race Law," *Washington Post*, July 17, 1913; "Race Policy Problem," *Washington Post*, September 30, 1913; "Race Issue up Again," *Washington Post*, March 7, 1914; "Upholds Race Purity," *Washington Post*, February 11, 1913; Green, *The Secret City*, 171–176; Moore, *Leading the Race*, 154–158. The university officially acknowledged a whites-only admission policy in 1921. Its readmission of blacks in 1936 was a first among other segregated city universities. C. Joseph Nuesse, "Segregation and Desegregation at the Catholic University of America," *Washington History* 9:1 (Spring/Summer 1997).
29. Moore, *Leading the Race*, 155, 158–159, 206–207; "Appeal to Wilson for Negro Clerks," *New York Times*, August 18, 1913; "Congressional Doings Told Briefly," *Washington Post*, January 7, 1915; Green, *The Secret City*, 176–177, 181, 191–192; Dennis R. Judd and Todd Swanstrom, *City Politics: The Political Economy of Urban America*, 5th ed. (New York: Pearson/Longman, 2006), 243; Blair A.

Ruble, *Washington's U Street: A Biography* (Washington, D.C.: Woodrow Wilson Center Press, 2010), 83–87.

30. The city's white population was also becoming increasingly Southern in origin. Carl Abbott, *Political Terrain: Washington, D.C., from Tidewater Town to Global Metropolis* (Chapel Hill: The University of North Carolina Press, 1999), 82–86; Crew, "Melding the Old and the New," 209. For more on the Great Migration, see Nicholas Lemann, *The Promised Land: The Great Migration and How It Changed America* (New York: Vintage Books, 1991); and Isabel Wilkerson, *The Warmth of Other Suns: The Epic Story of America's Great Migration* (New York: Random House, 2010).
31. Elizabeth Clark-Lewis, "'For a Real Better Life': Voices of African American Women Migrants, 1900–1930," in *Urban Odyssey*, 107–110; Green, *The Secret City*; Langston Hughes, "The Big Sea," in *Literary Washington, D.C.*, ed. Patrick Allen (San Antonio: Trinity University Press, 2012), 101, 103; Johnston, *Surviving Freedom*, 56–59; Audrey Elisa Kerr, *The Paper Bag Principle: Class, Colorism, and Rumor and the Case of Black Washington, D.C.* (Knoxville: University of Tennessee Press, 2006); Moore, *Leading the Race*, 10, 12, 31–32.
32. Quotes from Elizabeth Clark-Lewis, "'For a Real Better Life,'" 99, and by Clara Sharon Taylor and Loretta Carter Hanes in *Growing Up in Washington, D.C.: An Oral History*, ed. by Jill Connors (Charleston, SC: Arcadia Publishing, 2001), 31.
33. See, e.g., Connors, *Growing Up in Washington*, 128, 133. Georgetown became fashionable in the mid-1930s and, as a consequence, went from mostly black to racially mixed during the Great Depression (Green, *The Secret City*, 235–236).
34. Some blacks circumvented these restrictions by having a white person buy property on their behalf, as Mary Church Terrell and her husband did in the 1890s. Cherkasky, "'For Sale to Colored,'" 44, 45, 49, 50–51.
35. *Congressional Record,* April 15, 1926, p. 7494. Covenants had first been used in California in the 1890s to exclude the Chinese. Cherkasky, "'For Sale to Colored,'" 50. The Great Migration also contributed to the popularity of racial covenants in many Northern cities. See Judd and Swanstrom, *City Politics*, 243; Jon C. Teaford, *The Metropolitan Revolution* (New York: Columbia University Press, 2006), 23.
36. Crew, "Melding the Old and the New," 209–210, 220; Ruble, *Washington's U Street*, 52–54, 140–142. For more on the U Street neighborhood, see chapter 10.
37. Crew, "Melding the Old and the New," 215–216; Josef Eberle, "A German Editor's Impressions of 'The City Without a Mayor,'" *Washington Post,* June 5, 1949, emphasis added; Constance McLaughlin Green, *Washington, Vol. 2: Capital City, 1879–1950* (Princeton, NJ: Princeton University Press, 1963), 398; Louis Hyman (*Debtor Nation: The History of America in Red Ink,* Princeton: Princeton University Press, 2011), 63–66; Judd and Swanstrom, *City Politics*, 150–155; Teaford, *The Metropolitan Revolution*, 22–23. For more on the insufficient efforts of the federal government to help Washington blacks during the Great Depression, especially compared with whites, see Green, *The Secret City*, 232–235.

38. Crew, "Melding the Old and the New," 224; Michele F. Pacifico, "'Don't Buy Where You Can't Work': The New Negro Alliance of Washington," *Washington History* 6:1 (Spring/Summer 1994), 66–88. The theater stopped staging productions altogether until the early 1950s, claiming its lease prevented it from admitting African Americans. For this and other examples of local civil rights activism of the time, see Green, *The Secret City*, 290 and ch. 12.
39. Margaret E. Farrar, *Building the Body Politic: Power and Urban Space in Washington, D.C.* (Urbana: University of Illinois Press, 2008), 91–102; Howard Gillette Jr., *Between Justice and Beauty: Race, Planning, and the Failure of Urban Policy in Washington, D.C.* (Baltimore: Johns Hopkins University Press, 1995), 161–165; Harry S. Jaffe and Tom Sherwood, *Dream City: Race, Power, and the Decline of Washington, D.C.* (New York: Simon & Schuster, 1994), 29. It was followed four years later by passage of the National Housing Act, which encouraged the same kind of wholesale urban destruction in cities around the country.
40. Marvin Caplan, "Eat Anywhere!" *Washington History* 1:1 (Spring 1989); Steven J. Diner, "From Jim Crow to Home Rule," *The Washington Quarterly* 13:1 (New Year's 1989), 92; John Kelly, "Memories from the Front Lines of the Segregation Battle in the District," *Washington Post,* October 11, 2011; Nancy J. Weiss, *Farewell to the Party of Lincoln: Black Politics in the Age of FDR* (Princeton: Princeton University Press, 1983).
41. Green, *The Secret City*, 248–249, 286–288; quote from President's Committee on Civil Rights, *To Secure These Rights: The Report of the President's Committee on Civil Rights,* 1948, 95.
42. Bell Clement, "Pushback: The White Community's Dissent from *Bolling*," *Washington History* 16:2 (Fall/Winter 2004), 95; Diner, "From Jim Crow," 93; Green, *The Secret City*, 255–256; David A. Nichols, "'The Showpiece of Our Nation': Dwight D. Eisenhower and the Desegregation of the District of Columbia," *Washington History* 16 (Fall/Winter 2004), 51–55, 57–58; Robert Shogan, *Harry Truman and the Struggle for Racial Justice* (Lawrence: University of Kansas Press, 2013). Eisenhower was also the first president to hire an African American policy staffer (Lusane, *The Black History of the White House*, 271–277).
43. Judd and Swanstrom, *City Politics*, 243–244; Teaford, *The Metropolitan Revolution*, 62. Some Southern cities, such as Montgomery, Alabama, and Baton Rouge, Louisiana, had majority black populations briefly in the late nineteenth and early twentieth centuries. Gibson and Jung, *Historical Census Statistics for Large Cities,* tables 1 and 19.
44. Ed Bruske, "Shepherd Park: Activism Working within Tradition," *Washington Post,* August 15, 1987; Clement, "Pushback"; Teaford, *The Metropolitan Revolution*, 63–65.
45. See chapter 7.
46. Neil Spitzer, "A Secret City," *Wilson Quarterly* 13:1 (New Year's, 1989), 108.
47. Dennis E. Gale, *Washington, D.C.: Inner-City Revitalization and Minority Suburbanization* (Philadelphia: Temple University Press, 1987), 167; Hopkinson, *Go-Go Live*; Kip Lornell and Charles C. Stephenson, *The Beat!: Go-Go from*

Washington, D.C. (Jackson, MS: University Press of Mississippi, 2009); Ruble, *Washington's U Street*, 227; Brett Williams, *Upscaling Downtown: Stalled Gentrification in Washington, D.C.* (Ithaca: Cornell University Press, 1988).

48. Michael K. Fauntroy, *Home Rule or House Rule? Congress and the Erosion of Local Governance in the District of Columbia* (Dallas, TX: University Press of America, 2003), 137–138; Judd and Swanstrom, *City Politics*, 244–247; Nelson F. Kofie, 1999, *Race, Class, and the Struggle for Neighborhood in Washington, D.C.*, New York: Garland Publishing.
49. Jaffe and Sherwood, *Dream City*, 202, 209–212, 219–221, 223; Elliot Liebow 2003 [1967], *Tally's Corner: A Study of Negro Streetcorner Men* (Lanham: Rowman & Littlefield, 2003 [1967]), 41; see also Kofie, *Race*.
50. The problem of black homicide has remained a chronic and troubling one. In 2011, for instance, 90 percent of all those murdered in the district were African Americans. Metropolitan Police Department Annual Report for 2011, accessed January 2, 2013, http://mpdc.dc.gov/sites/default/files/dc/sites/mpdc/publication/attachments/ar_2011_0.pdf.
51. Sizable black suburbs can also be found outside such cities as Atlanta and Baltimore. Gale, *Washington, D.C.*, 113–114; Judd and Swanstrom, *City Politics*, 244–245; Karyn R. Lacy, *Blue-Chip Black: Race, Class, and Status in the New Black Middle Class* (Berkeley: University of California Press, 2007), 45.
52. Gale, *Washington, D.C.*, 168; quote from William P. O'Hare and William H. Frey, "Booming, Suburban, and Black," *Demographics Magazine* (September 1992), http://www.frey-demographer.org/briefs/B-1992-2_BecomingSuburban.pdf.
53. See chapter 9.
54. Hasia R. Diner and Steven J. Diner, "Washington's Jewish Community: Separate but Not Apart," in *Urban Odyssey*, 150; Mona E. Dingle, "*Gemeinschaft und Gemütlichkeit:* German American Community and Culture, 1850–1920," in *Urban Odyssey*, 116.
55. Frank Rich, "The De Facto Capital," *New York Times Magazine,* October 6, 2002.
56. Dingle, "*Gemeinschaft*," 115, 117; Margaret H. McAleer, "'The Green Streets of Washington': The Experience of Irish Mechanics in Antebellum Washington," in *Urban Odyssey*, 56.
57. Diner and Diner, "Washington's Jewish Community," 137–138; Howard Gillette, Jr., and Alan M. Kraut, "The Evolution of Washington's Italian American Community," in *Urban Odyssey,* 157, 160, 163–166. For more on Heurich, see chapter 9 and Dingle, "*Gemeinschaft*," 118–120. For more on the history of Jews in the Washington region, see David Altshuler, ed., *The Jews of Washington, D.C.: A Communal History Anthology* (Chappaqua, NY: Rossell Books, 1985).
58. Diner and Diner, "Washington's Jewish Community," 140, 144; Dingle, "*Gemeinschaft*," 124–125.
59. Dingle, "*Gemeinschaft*," 130–131; Diner and Diner, "Washington's Jewish Community," 136, 147.
60. See, e.g., Gillette, Jr. and Kraut, "The Evolution," 169. The D.C. branch of the JCC would later move into the old Sixteenth Street building. Diner and Diner, "Washington's Jewish Community," 148, 150–151.

61. For more on immigrants from elsewhere in Asia, see Beatrice Nied Hackett, "'We Must Become Part of the Larger American Family': Washington's Vietnamese, Cambodians, and Laotians," in *Urban Odyssey*, and Meeja Yu and Unyong Kim, "'We Came Here with Dreams': Koreans in the Nation's Capital," in *Urban Odyssey*.
62. Esther Ngan-ling Chow, "From Pennsylvania Avenue to H Street, NW: The Transformation of Washington's Chinatown," in *Urban Odyssey*, 191, 193–194.
63. Chow, "From Pennsylvania Avenue," 195.
64. A proposed convention center in the neighborhood was met with resistance; it was eventually built further west than originally proposed. Chow, "From Pennsylvania Avenue," 200–205.
65. Lydia DePillis, "More Disneyfication Coming to Chinatown," *Washington City Paper*, November 15, 2011; Eric M. Weiss, "Chiseling Away at Chinatown," *Washington Post*, February 14, 2005.
66. Lisa Benton-Short and Marie Price, introduction to *Migrants to the Metropolis: The Rise of Immigrant Gateway Cities*, ed. by Marie Price and Lisa Benton-Short (Syracuse, NY: Syracuse University Press, 2008), 20; Elizabeth Chacko, "Washington D.C.: From Biracial City to Multiethnic Gateway," in *Migrants to the Metropolis*, 209; Walter Nicholls, "A New Chinatown," *Washington Post*, October 22, 2003; Bonnie Tsui, 2011, "The End of Chinatown," *The Atlantic*, October 28.
67. Olivia Cadaval, "The Latino Community: Creating an Identity in the Nation's Capital," in *Urban Odyssey*, 233–235. Salvadorans consist of about 14 percent, or 167,000, of all foreign-born in the area. Cadaval, "The Latino Community," 242; Audrey Singer, "Metropolitan Washington: A New Immigrant Gateway," in *Hispanic Migration and Urban Development: Studies from Washington DC*, ed. by Enrique S. Pumar (Bingley, UK: Emerald Press, 2012), 13–14.
68. Cadaval, "The Latino Community," 245–246; Emily Friedman, "Mount Pleasant Riots: May 5 Woven into Neighborhood's History," *WAMU*, May 5, 2011.
69. Cadaval, "The Latino Community," 243; Mark Hugo Lopez and Daniel Dockterman, "A Growing and Diverse Population: Latinos in the Washington, DC Metropolitan Area," in *Hispanic Migration*, 92–98; Brigid Schulte, "Wheaton Seeks Bridge across Cultures," *Washington Post*, February 15, 2011.
70. As of 2000. Chacko, "Washington D.C.," 210.
71. One common estimate is that 200,000 Ethiopians reside in the D.C. metropolitan area. Samantha Friedman, Audrey Singer, Marie Price, and Ivan Cheung, "Race, Immigrants, and Residence: A New Racial Geography of Washington, D.C.," *The Geographical Review* 95:2 (2005), 211; Brian Westley, "Washington: Nation's Largest Ethiopian Community Carves Niche," *Associated Press*, October 17, 2005.
72. Chacko, "Washington D.C.," 217, 221–222; Trymaine Lee, "As Black Population Declines in Washington, D.C., Little Ethiopia Thrives," *Huffington Post*, April 8, 2011, accessed November 18, 2012, http://www.huffingtonpost.com/2011/04/08/black-population-declies-dc-little-ethiopia-thrives_n_846817.html; Terrence Lyons, "The Ethiopian Diaspora and Homeland Conflict," *Institute*

for Conflict Analysis and Resolution (George Mason University, 2009), 592–593, http://portal.svt.ntnu.no/sites/ices16/Proceedings/Volume%202/Terrence%20Lyons-%20The%20Ethiopian%20Diaspora%20and%20Homeland%20Conflict.pdf; Walter Nicholls, "Washington's Little Ethiopia," *Washington Post*, May 18, 2005; Bereket H. Selassie, "Washington's New African Immigrants," in *Urban Odyssey*, 264; Brian Westley, "Washington."

73. Lyons, 2009; Annys Shin, "Ethiopian Soccer Tournament Promoting Unity Leads to Division," *Washington Post*, July 5, 2012; Emily Wax, "Ethiopian Yellow Pages: Life, by the Book," *Washington Post*, June 8, 2011. See also Wilbur Zelinsky and Barrett A. Lee, "Heterolocalism: An Alternative Model of the Sociospatial Behavior of Immigrant Ethnic Communities," *International Journal of Population Geography* 4:4 (1998).
74. Lisa Benton-Short and Marie Price, introduction to *Migrants to the Metropolis*, 2, 6, 10, 16; Karen Destorel Brown, "Expanding Affordable Housing through Inclusionary Zoning: Lessons from the Washington Metropolitan Area," Center on Urban and Metropolitan Policy (Washington, D.C.: Brookings Institution 2001); Chacko, "Washington D.C.," 213; William H. Frey, Brookings Institution and University of Michigan Social Science Data Analysis Network's analysis of 2005–2009 American Community Survey and 2000 Census Decennial Census tract data, http://www.psc.isr.umich.edu/dis/census/segregation.html, accessed November 15, 2012; Carol Morello and Dan Keating, "The New American Neighborhood," *Washington Post*, October 30, 2011; Myron Orfield and Thomas Luce, "America's Racially Diverse Suburbs: Opportunities and Challenges," Institute of Metropolitan Opportunity, University of Minnesota Law School, July 20, 2012, 37–38; Audrey Singer, "The Rise of New Immigrant Gateways," *The Living Cities Census Series* (Washington, D.C.: Brookings Institution, February 2004); Friedman et al., Race, Immigrants, and Residence"; Jill H. Wilson and Audrey Singer, *State of Metropolitan America* (Washington, D.C.: The Brookings Institution, October 13, 2011). The trend toward greater suburban diversity is seen elsewhere in the United States too (Judd and Swanstrom, *City Politics*, 249, 256). The term *ethnoburb* comes from Wei Li, "Anatomy of a New Ethnic Settlement: The Chinese Ethnoburb in Los Angeles," *Urban Studies* 35:3 (1998).
75. William H. Frey, "Millennial and Senior Migrants Follow Different Post-Recession Paths," Brookings Institution Report, November 15, 2013; Haya El Nasser, "Young and Educated Show Preference for Urban Living," *USA Today*, April 1, 2011; Carol Morello, Dan Keating, and Steve Hendrix, "Capital Hip: D.C. Is Getting Younger," *Washington Post*, May 5, 2011.

Chapter 9

1. Robert Samuels, "Community Deluged by Sewage," *Washington Post*, August 26, 2012.
2. Dennis R. Judd and Todd Swanstrom, *City Politics: The Political Economy of Urban America*, 5th ed. (New York: Pearson/Longman, 2006), 3.

3. Joel Achenbach, *The Grand Idea: George Washington's Potomac and the Race to the West* (New York: Simon and Schuster, 2004), esp. 122–123, 127, 213, 267; Kenneth R. Bowling, *The Creation of Washington, D.C.: The Idea and Location of the American Capital* (Fairfax, VA: George Mason University Press, 1991).
4. See, e.g., Michael Dear and Nicholas Dahmann, "Urban Politics and the Los Angeles School of Urbanism," in *The City, Revisited: Urban Theory from Chicago, Los Angeles, and New York*, edited by Dennis R. Judd and Dick Simpson (Minneapolis: University of Minnesota Press, 2011), 69–70.
5. Scott W. Berg, *Grand Avenues: The Story of the French Visionary Who Designed Washington, D.C.* (New York: Vintage Books, 2007), 104–105. For more on the importance of rivers and commerce in location, see Carl Abbott, *Political Terrain: Washington, D.C., from Tidewater Town to Global Metropolis* (Chapel Hill: The University of North Carolina Press, 1999).
6. Achenbach, *The Grand Idea*, 241–242; Berg, *Grand Avenues*, 128–129.
7. The canal was intended to get as far as the Ohio River but never did. Achenbach, *The Grand Idea*, 215–216, 244, 262; Constance McLaughlin Green, *Washington, Vol. 1: Village and Capital, 1800–1878* (Princeton, NJ: Princeton University Press, 1962), 156–157, 193–194. Early efforts in Congress to build a network of roads from the district also faltered (e.g., Abbott, *Political Terrain*, 35–36).
8. James H. S. McGregor, *Washington from the Ground Up* (Cambridge, MA: Harvard University Press, 2007), 265. Because Georgetown and Washington had separate governments and budgets through the mid-1800s, they "pursued prosperity individually." Alan Lessoff, *The Nation and Its City: Politics, "Corruption," and Progress in Washington, D.C., 1861–1902* (Baltimore, MD: Johns Hopkins University Press, 1994), 31.
9. Kenneth R. Bowling, "From 'Federal Town' to 'National Capital': Ulysses S. Grant and the Reconstruction of Washington, D.C.," *Washington History* 14:1 (2002), 14–18; Constance McLaughlin Green, *Washington, Vol. 2: Capital City, 1879–1950* (Princeton, NJ: Princeton University Press, 1963), 12–13; Robert Harrison, *Washington during Civil War and Reconstruction: Race and Radicalism* (New York: Cambridge University Press, 2011), 58, 156–157; Kathryn Allamong Jacob, *Capital Elites: High Society in Washington, D.C. after the Civil War* (Washington, D.C.: Smithsonian Institution Press, 1995), 140–150; Lessoff, *The Nation and Its City*, chs. 2–3. The Heurich House museum is a good source for information about Heurich and his Dupont Circle mansion (http://www.heurichhouse.org/history/).
10. James M. Goode, *Best Addresses: A Century of Washington's Distinguished Apartment Houses* (Washington, D.C.: Smithsonian Institution Press, 1988), 50; Lessoff, *The Nation and Its City*, 9, 159–162. For more on politicians investing in D.C., see chapter 4.
11. Abbott, *Political Terrain*, 102–103 and table 4.2; see also chapter 1.
12. Abbott, *Political Terrain*, 101, 120; quote from Goode, *Best Addresses*, 52. For more on Cafritz, see Burt Solomon, *The Washington Century: Three Families and the Shaping of the Nation's Capital* (New York: HarperCollins, 2004).

13. This continues to be an issue with the destruction of older public housing stock in the city, documented for instance in the local documentary film *Chocolate City* (http://dcist.com/2008/01/25/chocolate_city.php).
14. Judd and Swanstrom, *City Politics*, ch. 6; Walter A. Scheiber, "Washington's Regional Development," *Records of the Columbia Historical Society, Washington, D.C.* 49 (1973–1974); quote from Sam Bass Warner, Jr., *The Urban Wilderness: A History of the American City* (Berkeley: University of California Press, 1995[1972]), 123.
15. Kenneth T. Jackson, *Crabgrass Frontier: The Suburbanization of the United States* (New York: Oxford University Press, 1985), 266; Judd and Swanstrom, *City Politics*, ch. 6; see also chapter 8.
16. Quote from Jon C. Teaford, *The Metropolitan Revolution* (New York: Columbia University Press, 2006), 127. The decline of American cities after World War II has been written about by many. See, for instance, Warner, *The Urban Wilderness*; Robert M. Fogelson, *Downtown: Its Rise and Fall, 1880–1950* (New Haven: Yale University Press, 2003); Douglas W. Rae, *City: Urbanism and Its End* (New Haven: Yale University Press, 2005); and Thomas J. Sugrue, *Origins of the Urban Crisis: Race and Inequality in Postwar Detroit* (Princeton, NJ: Princeton University Press, 1996).
17. Judd and Swanstrom, *City Politics*, 190–194; Teaford, *The Metropolitan Revolution*, 127, quote p. 164.
18. Campbell Gibson and Kay Jung, *Historical Census Statistics on Population Totals by Race, 1790 to 1990, and by Hispanic Origin, 1970 to 1990, for Large Cities and Other Urban Places in the United States*, Working Paper No. 76, U.S. Census Bureau (February 2005); "DC vs. VA vs. MD," *Washingtonian Magazine*, October 2012, 75.
19. Teaford, *The Metropolitan Revolution*, 171, quote p. 169.
20. Tyson Freeman, "The 1980s: (Too) Easy Money Fuels a Building Boom," *National Real Estate Investor*, September 30, 1999 (http://nreionline.com/mag/real_estate_easy_money_fuels/); Robert G. Kaiser, "Big Money Created a New Capital City," *Washington Post*, April 8, 2007. A brief summary of Washington's downtown development in the 1980s can be found in Stephen J. McGovern, *The Politics of Downtown Development: Dynamic Political Cultures in San Francisco and Washington, D.C.* (Lexington: University Press of Kentucky, 1998), 197–210. For more on the convention center, see chapter 7.
21. See, e.g., Judd and Swanstrom, *City Politics*, 249; McGovern, *The Politics of Downtown Development*, 202–208. On the urban renewal movement more generally, see Judd and Swanstrom, *City Politics*, ch. 13 and Teaford, *The Metropolitan Revolution*.
22. Judd and Swanstrom, *City Politics*, 256, 362. For more on D.C.'s economic revitalization in the 1990s, see Gerry Widdicombe, "The Fall and Rise of Downtown D.C.," *Urbanist*, January 2010 (http://www.spur.org/publications/library/article/fall_and_rise_downtown_dc).
23. Annie Lowrey, "Washington's Economic Boom, Financed by You," *New York Times*, January 10, 2013; U.S. Census Bureau, *Patterns of Metropolitan and*

Micropolitan Population Change: 2000 to 2010, September 2012, http://www.census.gov/prod/cen2010/reports/c2010sr-01.pdf, table 3.7.

24. *Citizens Financial Report 2011,* Office of Chief Financial Officer 2011, District of Columbia: Washington, D.C., p. 8. D.C. is but one of many American cities that have looked to sports, conventions, and even gambling as a way to bring more visitors and money within its borders; see Judd and Swanstrom, *City Politics*, 372–386.
25. Judd and Swanstrom, *City Politics*, 3. The other two are politics of governance (how to address citizens' concerns through the democratic process) and the politics of defended space (dealing with separations along racial, ethnic, and economic lines).
26. Greater Washington Board of Trade, *Greater Washington 2010 Regional Report*, Washington, D.C., 4, 6, 20, and 27; PricewaterhouseCoopers, 2009, *UK Economic Outlook* (November), table 3.3.
27. Richard Florida, "The 25 Most Economically Powerful Cities in the World," *The Atlantic*, September 15, 2011, accessed December 8, 2013, http://www.theatlanticcities.com/jobs-and-economy/2011/09/25-most-economically-powerful-cities-world/109/#slide9.
28. As of September 2011. *Comprehensive Annual Financial Report for Fiscal Year 2011,* Office of the Chief Financial Officer, District of Columbia, 8.
29. Greater Washington Board of Trade, *Greater Washington 2010 Regional Report*, 11, 24; Kaiser, "Big Money"; "Washington Post 200," *Washington Post* (2010), accessed December 9, 2013 (http://www.washingtonpost.com/wp-srv/special/business/post200-2009/post-200-graphic.html). Another report from 2011 counted 172 former lawmakers who are "working on behalf of private interests in Washington's influence-peddling industry." Justin Elliott and Zachary Roth, "Shadow Congress: More than 170 Former Lawmakers Ply the Corridors of Power as Lobbyists," *Talking Points Memo,* June 1, 2010, accessed December 3, 2013, http://tpmmuckraker.talkingpointsmemo.com/2010/06/shadow_congress_former_lawmakers_become_lobbyists.php.
30. Danielle Douglas, "Uneven Upswing for D.C. Area Tourism," *Washington Post,* May 14, 2012; Glenn Fowler, "David Lloyd Kreeger Dead at 81; Insurance Official and Arts Patron," *New York Times*, November 20, 1990; Green, *Washington, Vol. 1*, 84.
31. For more on the features and advantages of an agglomeration economy, see Brendan O'Flaherty, *City Economics* (Cambridge, MA: Harvard University Press, 2005), 16–25.
32. Abbott, *Political Terrain*, 77–81. The board was politically quite influential for many decades; see chapter 7. Recognizing the growth of business interests outside of D.C. proper, the board later renamed itself the Greater Washington Board of Trade.
33. Abbott, *Political Terrain*, 112–119.
34. For instance, the state of Virginia offered cash grants to successfully convince Volkswagen to put their North American headquarters in Herndon, outside of

D.C. Zachary A. Goldfarb, "Volkswagen Moving to Herndon," *Washington Post,* September 6, 2007.

35. For histories of the NoMa neighborhood's recent growth, see Rachel MacCleery and Jonathan Tarr, "NoMa: The Neighborhood That Transit Built," *Urbanland,* February 29, 2012, accessed December 9, 2013, http://urbanland.uli.org/Articles/2012/Jan/MacCleeryNOMA; and Rick Rybeck, "Using Value Capture to Finance Infrastructure and Encourage Compact Development," *Public Works Management and Policy* 8:4 (2004).
36. Philip M. Dearborn and Stephanie Richardson, "Home Buyer Credit Widely Used," Research Report, *Greater Washington Research Center,* May 5, 1999, 3. On H Street, N.E. see Oramenta Newsome and Michael Rubinger, "The H Street Revival: Not a Miracle, Just Community Development," *Washington Post,* March 24, 2013. Leadership has mattered for the revitalization of other city centers as well. See, e.g., Costas Spirou, "Both Center and Periphery: Chicago's Metropolitan Expansion and the New Downtowns," in *The City, Revisited.*
37. Jon Bouker et al., "Building the Best Capital City in the World," Research Report (Washington, D.C.: Brookings Institution, 2008), 160.
38. In addition, the district's budget cannot be implemented if Congress does not enact the district appropriations bill, though a referendum enacted in 2013 was designed to decouple the local government's budget from Congress. See chapter 7.
39. Jonathan O'Connell, "St. Elizabeths Renovation as Security Campus Faces Resistance," *Washington Post,* March 30, 2012; Josh Hicks, "How Much Did Closing the Government Cost?" *Washington Post,* October 21, 2013; Corinne Reilly, "Fairfax Already Feeling Chill of Cuts," *Washington Post,* October 24, 2012; quote from Marc Fisher, "District No Joke Now," *Washington Post,* October 26, 2011; see also Lowrey, "Washington's Economic Boom."
40. Ross Douthat, "Washington versus America," *New York Times,* September 25, 2012.
41. See Andrew Cohen, "There's America—and Then There's Washington," *The Atlantic*, October 27, 2011, accessed October 28, 2011, http://www.theatlantic.com/politics/archive/2011/10/theres-america-and-then-theres-washington/247442/.
42. *Comprehensive Annual Financial Report for Fiscal Year 2011,* p. 8. In a more recent news article, economist Stephen Fuller has estimated the percentage of the regional economy dependent on the U.S. government at closer to 40 percent. Lowrey, "Washington's Economic Boom."
43. The universities were Georgetown, George Washington, Howard, American, and Catholic. The four hospitals were Washington Hospital Center, Children's, Georgetown, and Providence. Fannie Mae was the sole entity on the list that was neither a hospital nor a university. Michael Neibauer, "Howard University No Longer D.C.'s Top Employer," *Washington Business Journal,* February 4, 2011.
44. Forrest McDonald, *Novus Ordo Seclorum* (Lawrence, KS: University Press of Kansas, 1985), 190–191. Washington and Madison both called for a university in the district during their presidencies; see George Washington's Eighth Annual

Address to Congress, December 7, 1796, and James Madison's Eighth Annual Address to Congress, December 3, 1816.

45. Abbott, *Political Terrain*, 105–106, 109–112, 121–122; Stephen S. Fuller, "The Impact of the Consortium of Universities of the Washington Metropolitan Area on the Economies of the Washington Metropolitan Area and District of Columbia," May 2011, http://cra.gmu.edu/pdfs/researach_reports/recent_reports/Economic_Impacts_of_Washington_Consortium_Universities.pdf. Congress did issue a charter for Georgetown University in 1815, the first university so designated, allowing it to confer degrees.
46. Abbott, *Political Terrain*, 123–125, quote p. 123. According to Florida, the D.C. area is home to a major concentration of this class and is expected to see some of the fastest growth in jobs nationally, both for the creative class and in general. Richard Florida, "Where the Creative Class Jobs Will Be," *The Atlantic.com*, August 25, 2010, accessed December 8, 2013, http://www.theatlantic.com/business/archive/2010/08/where-the-creative-class-jobs-will-be/61468//; and "Where the Jobs Will Be," *The Atlantic.com*, August 17, 2010, accessed December 7, 2013, http://www.theatlantic.com/business/archive/2010/08/where-the-jobs-will-be/61459/.
47. Danielle Douglas, "Can the Private Sector Save Washington?" *Washington Post Capital Business Section*, December 19, 2011; Goldfarb, "Volkswagen Moving"; Lowrey, "Washington's Economic Boom."
48. The *Washington Post* keeps track of the area's largest employers in their annual "Washington Post 200" list. See http://www.washingtonpost.com/post200/.
49. "Washington Post 200," *The Washington Post* (http://www.washingtonpost.com/wp-srv/special/business/post200-2009/post-200-graphic.html).
50. Mark Abrahamson, *Global Cities* (New York: Oxford University Press, 2004), 4–5; Fisher, "District No Joke Now"; Greater Washington Board of Trade, *Greater Washington 2010 Regional Report*, 21.
51. Washington has been a tourism city since the 1910s (Green, *Washington, Vol. 2*, 175). Domestic tourism brings in far more dollars than international tourism. In 2011, for instance, 1.8 million foreigners visited Washington, compared with 16.1 million American tourists; the latter spent an estimated $6 billion dollars in the city. Douglas, "Uneven Upswing"; "Washington DC's 2011 Visitor Statistics," *Destination DC*, 2011, accessed February 11, 2013, http://planning.washington.org/images/pdfs/2011_VisitorStatistics2.pdf. Tourism is also a highly desired source of revenue for other cities; see Judd and Swanstrom, *City Politics*, 372–373.
52. Jon Bouker et al., "Building the Best Capital City," 23–25.
53. On the economics of building height restrictions, see O'Flaherty, *City Economics*, 130.
54. Paul Ceruzzi, *Internet Alley: High Technology in Tysons Corner, 1945–2005* (Cambridge: MIT Press, 2008); Stephen S. Fuller, "Northern Virginia's Economic Transformation," November 2011, http://cra.gmu.edu/pdfs/studies_reports_presentations/By_The_Numbers_NoVa_Drives_Area_Growth.pdf; "Great Seneca

Science Corridor," Maryland-National Capital Park and Planning Commission, accessed January 8, 2013, http://www.montgomeryplanning.org/community/gaithersburg/index.shtm.

55. Center on Urban and Metropolitan Policy, "A Region Divided: The State of Growth in Greater Washington, D.C.," Research Report (Washington, D.C.: Brookings Institution, 1999), 3, 11; Bureau of Labor Statistics, February 2013 unemployment rate (not seasonably adjusted), http://www.bls.gov/ro3/mdlaus.htm, http://www.bls.gov/ro3/valaus.htm, accessed April 30, 2013; Sarah Halzack, "D.C. Area Bucks Trend of 'Job Sprawl,'" *Washington Post*, April 22, 2013.
56. They were Wards 5 through 8. "District of Columbia Labor Force, Employment, Unemployment, and Unemployment by Ward," Department of Employment Services (District of Columbia, February 2013), accessed April 30, 2013, http://does.dc.gov/sites/default/files/dc/sites/does/page_content/attachments/DC%20Ward%20Data%20Feb13-Jan13-Feb12.pdf.
57. A similar problem emerged during the city's development boom of the 1980s; see McGovern, *The Politics of Downtown Development*, 202.
58. The city's homeless population grew significantly in 2011 yet dropped in Maryland and Virginia, though it is unclear whether this is due to in-migration of the homeless. Ivan V. Natividad, "A Haven for the Homeless," *Roll Call,* December 14, 2011.
59. Caitlin Biegler, "A Big Gap: Income Inequality in the District Remains One of the Highest in the Nation," *DC Fiscal Policy Institute, Center on Budget and Policy Priorities,* March 8, 2012, http://www.dcfpi.org/wp-content/uploads/2012/03/03-08-12incomeinequality1.pdf; Timothy R. Homan, "Unemployment Rate in Washington's Ward 8 Is Highest in U.S," *Bloomberg News,* March 30, 2011.
60. Howard Gillette, Jr., *Between Justice and Beauty: Race, Planning, and the Failure of Urban Policy in Washington, D.C.* (Baltimore: Johns Hopkins University Press, 1995); McGovern, *The Politics of Downtown Development*, 204–206 and chs. 10–11; Teaford, *The Metropolitan Revolution*, 115–116; Emily Wax, "Black Middle Class Is Redefining Anacostia," *The Washington Post,* July 29, 2011; quote from Marc Fisher, "Does Culture Follow the Census?" Washington Post, April 11, 2011. See also Natalie Hopkinson, *Go-Go Live: The Musical Life and Death of Chocolate City* (Durham: Duke University Press, 2012), xi–xii.
61. See, e.g., McGovern, *The Politics of Downtown Development*, 197–198; Garance Franke-Ruta, "Facts and Fictions of D.C.'s Gentrification," *The Atlantic Cities,* August 10, 2012, http://www.theatlanticcities.com/politics/2012/08/facts-and-fictions-gentrification-dc/2914/.
62. Quotes from Wax, "Black Middle Class" and Shani O. Hilton, "Confessions of a Black Gentrifier," *Washington City Paper,* March 18, 2011.
63. O'Flaherty, *City Economics*, 531–540, 545–549; "A Whole New Ballpark," *Pacific Standard*, November/December 2012, 12–13.
64. See, for example, Peter Hermann, "As D.C. Changes, Police Try to Adapt," *Washington Post,* August 5, 2012.

65. Ashley Halsey, III, "We're No. 1 in Traffic Gridlock," *Washington Post,* September 27, 2011. Traffic problems have beset northern Virginia for years; see, e.g., Teaford, *The Metropolitan Revolution*, 200–201. On the slugging movement, see Emily Badger, "The People's Transit," *Miller-McCune*, March/April 2011, 57–65.
66. Judd and Swanstrom, *City Politics*, 427.

Chapter 10

1. Many religious orders purchased or rented houses in the Brookland area. Other major Catholic institutions in Brookland include the Dominican House of Studies, Trinity University, the Pope John Paul II Cultural Center and the Ukrainian Catholic National Shrine.
2. As in many other cities, there is some inconsistency and ambiguity in defining clear neighborhood boundaries and, as a result, how many individual neighborhoods exist within the District of Columbia. For instance, the city's Office of Planning divides the District of Columbia into 39 neighborhood clusters and recognizes 131 separate neighborhoods, whereas the D.C. Office of Tax and Revenue divides the district into 68 individual neighborhoods. For detailed information about the neighborhood clusters and individual neighborhoods, see Peter A. Tatian et al, *State of Washington, D.C.'s Neighborhoods* (Washington, D.C.: The Urban Institute, 2008), 4.
3. Carl Abbott, *Political Terrain: Washington, D.C., from Tidewater Town to Global Metropolis* (Chapel Hill: University of North Carolina Press, 1999). As he considers how Washington, D.C., emerged in a regional borderland and how competing groups, agendas, and needs have influenced the development of the city, Abbott argues that the "assumption of rootless residents implies that Washington cannot be understood as an identifiable place, but only as a collection of place seekers" (4).
4. Richard Nixon quoted in Frank DeFord, "Home without Homeowners: Washington, D.C.," *Sports Illustrated* (July 2, 1979): 68. See Abbott, *Political Terrain*, 3.
5. Abbott, *Political Terrain*, 134; Tatian, *State of Washington, D.C.'s Neighborhoods*, 4; U.S. Census Bureau, http://2010.census.gov/2010census/data/.
6. Kathryn Schneider Smith, ed., *Washington at Home: An Illustrated History of Neighborhoods in the Nation's Capital*, second edition (Baltimore: The Johns Hopkins University Press, 2010), 3. L'Enfant's plan is discussed in more detail in chapter 1. While large sections of the city reflect the basic street plans as laid out by L'Enfant, some neighborhoods—namely, those developed later, and thus outside L'Enfant's original plan—have distinct street arrangements. See Kenneth Jackson, *Crabgrass Frontier: The Suburbanization of the United States* (New York: Oxford University Press, 1985), 74.
7. *Places and Persons on Capitol Hill: Stories and Pictures of a Neighborhood* (Washington, D.C.: Capitol Hill Southeast Citizens Association, 1960), 10; James

Sterling Young, *The Washington Community, 1800–1828* (New York: Columbia University Press, 1966).

8. For more information about the history of Capitol Hill, see *The Ruth Ann Overbeck Capitol Hill History Project*, http://capitolhillhistory.org/index.html.
9. Samuel C. Busey, *Personal Reminiscences and Recollections of Forty-Six Years' Membership in the Medical Society of the District of Columbia, and Residence in This City, with Biographical Sketches of Many of the Deceased Members* (Philadelphia: Dornan, 1895), 71, 83.
10. Richard P. Greene and James B. Pick, *Exploring the Urban Community: A GIS Approach*, second edition (Boston: Prentice Hall, 2006), 141.
11. According to the Greenbelt Museum website, Greenbelt, MD was "designed and built as one of three 'green belt towns' during the Great Depression," and it was constructed to "provide work relief for the unemployed, provide affordable housing for low income workers, and be a model for future town planning in America" and balance the best features of rural and urban life (*Greenbelt Museum*, http://greenbeltmuseum.org/history). For more, see Susan L. Klaus, *Links in the Chain: Greenbelt, Maryland, and the New Town Movement in America* (Washington, D.C.: George Washington University Press, 1987); Cathy D. Knepper, *Greenbelt, Maryland: A Living Legacy of the New Deal* (Baltimore: Johns Hopkins University Press, 2001); Zane L. Miller, *The New Deal in the Suburbs: A History of the Greenbelt Town Program, 1935–1954* (Columbus: Ohio State University Press, 1971); and Mary Lou Williamson, ed., *Greenbelt: History of a New Town, 1937–1987* (Norfolk, VA: Dunning, 1987).
12. Jackson, *Crabgrass Frontier*, 39.
13. For more information about the Historic Mount Pleasant Neighborhood, see *The Mount Pleasant Main Street Organization*, http://www.mtpleasantdc.org; "Mount Pleasant Historic District," *National Park Service National Register of Historic Places*, http://www.cr.nps.gov/nr/travel/wash/dc96.htm; and Linda Low and Mara Charkasky, "Mount Pleasant" in Smith, *Washington at Home*, 213–227.
14. For a brief history of the development of Chevy Chase, see Jackson, *Crabgrass Frontier*, 122–124.
15. As Judd and Swanstrom point out, the automobile allowed developers to "fill in the gaps" of undeveloped land along tracks. With the expanded possibilities for development, "vast new tracts of land were opened to land speculation and suburban development, and the upper-middle class invested much of its newfound money in suburban real estate." Dennis R. Judd and Todd Swanstrom, *City Politics: The Political Economy of Urban America*, eighth edition (Boston: Longman, 2012), 157.
16. Joel Garreau, *Edge City: Life on the New Frontier* (New York: Doubleday, 1991).
17. See Garreau, *Edge City*, 303–422 for more information about how he sees Tysons Corner, Virginia, as a typical edge city. Detailed information about the development of Tysons Corner can be found in Paul Ceruzzi, *Internet Alley: High Technology in Tysons Corner, 1945–2005* (Boston: MIT Press, 2008).

18. Jonathan O'Connell, "Can City Life Be Exported to the Suburbs?" *Washington Post*, September 7, 2012; Corinne Riley and Victor Zapana, "Tysons Corner Is Unofficially Dropping the 'Corner' from Its Name," *Washington Post*, October 4, 2012.
19. Jonathan O'Connell, "Tysons Corner: The Building of an American City," *Washington Post*, September 24, 2011.
20. For more details about the planning and construction of a highway system in the New York City metropolitan region, see Robert Caro, *The Power Broker: Robert Moses and the Fall of New York* (New York: Alfred A. Knopf, 1974).
21. Jeremy Knorr, "Political Parameters: Finding a Route for the Capital Beltway, 1950–1964," *Washington History* 19/20 (2007/2008): 12–13; *Transportation Plan; National Capital Region. The Mass Transportation Survey Report, 1959* (Washington, D.C.: National Capital Planning Commission, 1959). For a more detailed explanation of the proposed road plans and the reactions from citizens who would be affected, see Bob Levey and Jane Freundel Levey, "End of the Roads." *Washington Post*, November 26, 2000. The I-495 Capital Beltway was officially completed in August 1964.
22. See Zachary Schrag, *The Great Society Subway: A History of the Washington Metro* (Baltimore: The Johns Hopkins University Press, 2006) for details about the Washington Metro system.
23. William H. Lucy and David L. Phillips, *Tomorrow's Cities, Tomorrow's Suburbs* (Washington, D.C.: Planners Press, 2006), 317–319; Schrag, *The Great Society Subway*, 2.
24. Lucy, *Tomorrow's Cities, Tomorrow's Suburbs*, 317–319.
25. Constance McLaughlin Green, *The Secret City: A History of Race Relations in the Nation's Capital* (Princeton: Princeton University Press, 1967). This phenomenon is also described by Sandra Fitzpatrick and Maria Goodwin: "Forced to overcome the barriers imposed by segregation and discrimination, black Washingtonians created a vital social and economic culture within their carefully delineated neighborhoods." Sandra Fitzpatrick and Maria R. Goodwin, *The Guide to Black Washington: Places and Evens of Historical and Cultural Significance in the Nation's Capita*l, revised illustrated edition (New York: Hippocrene Books, 2001), 13. See also chapter 8.
26. For more information about U Street, see Blair A. Ruble, *Washington's U Street: A Biography* (Baltimore: Johns Hopkins University Press, 2010); Paul A. Williams, *Greater U Street* (Charleston, SC: Arcadia Publishing, 2001); and "The Greater U Street Historic District," National Park Service, http://www.nps.gov/nr/travel/wash/dc63.htm. For more information about other African American neighborhoods within Washington, D.C., see Fitzpatrick and Goodwin, *The Guide to Black Washington* and "The African American Heritage Trail in Washington, D.C.," *Cultural Tourism DC*, http://www.culturaltourismdc.org/things-do-see/tours-trails/african-american-heritage-trail-washington-dc.
27. The documentary was created in 1989 as a part of Georgetown University's bicentennial celebration, and the book followed that project. Both shared the

same purpose: to preserve the heritage of the black community of Georgetown and recover an often overlooked aspect of the neighborhood's history. See *Black Georgetown Remembered: A Documentary Video on the History of the Black Georgetown Community*, VHS, directed by David W. Powell (Washington, D.C.: Georgetown University, 1989) and Kathleen M. Lesko et al, *Black Georgetown Remembered: A History of Its Black Community from the Founding of 'The Town of George' in 1751 to the Present Day* (Washington, D.C.: Georgetown University Press, 1991).

28. Lessoff 1994, 205–208; Smith, *Washington at Home*, 11–12. For a more detailed description of how the local political situation in Washington, D.C., led to the creation of powerful community action groups, see George W. McDaniel and John N. Pearce, eds., *Images of Brookland: The History and Architecture of a Washington Suburb*, revised and enlarged by Martin Aurand (Washington, D.C.: Center for Washington Area Studies, George Washington University, 1982).
29. McDaniel and Pearce, *Images of Brookland*, 30–31.
30. Smith, *Washington at Home*, 12.
31. For instance, the city boasts a Neighborhood Heritage Trails project, organized and maintained by a group known as Cultural Tourism DC. Working with neighborhood and local historical associations, Cultural Tourism DC developed nineteen self-guided walking routes marked by illustrated signs to celebrate the unique history and cultural heritage of different historic neighborhoods throughout Washington, D.C. (see http://www.culturaltourismdc.org/ for more). Other communities in the greater metropolitan area have created similar walking tours, such as the Silver Spring Heritage Trail, part of the larger Montgomery County Main Street Heritage Trail Project (see http://www.montgomerycountymd.gov/content/council/mem/ervin_v/pdfs/ervin-silverspringhistorytrail41210.pdf for more).
32. See "Advisory Neighborhood Commissions," *District of Columbia*, http://anc.dc.gov/.
33. "The Voices on 14th Street," a series of interviews sponsored by the Humanities Council of Washington, examines the vibrant civic activism in the Columbia Heights area following the 1968 race riots. For transcripts, see http://www.wdchumanities.org/dcdm. The Hillcrest Community Civic Organization completed an oral history project in 2010, focusing on the history of the neighborhood and asking residents how they would define a strong neighborhood in Washington, D.C. See Michelle Phipps-Evans, "What Makes a Strong DC Community—A Hillcrest Community Civic Association Oral History Project—Overview," *DC Digital Museum*, http://www.wdchumanities.org/dcdm/items/show/1518.

Bibliography

"A 20-Cent Attitude." *Washington Post,* July 7, 1935.

"A Whole New Ballpark." *Pacific Standard* (November/December 2012): 12–13.

Abbott, Carl. *Political Terrain: Washington, D.C., from Tidewater Town to Global Metropolis.* Chapel Hill: The University of North Carolina Press, 1999.

Abrahamson, Mark. *Global Cities.* New York: Oxford University Press, 2004.

Achenbach, Joel. *The Grand Idea: George Washington's Potomac and the Race to the West,* New York: Simon and Schuster, 2004.

Adam, Robert. *Classical Architecture: A Comprehensive Handbook to the Tradition of Classical Style.* New York: Harry N. Abrams, Inc., Publishers, 1990.

Adams, Henry. *Democracy.* New York: Harmony Books, 1880 [1981].

Allen, Theodore W. *The Invention of the White Race: The Origin of Racial Oppression in Anglo America.* New York: Verso, 1994.

Allen, Thomas B. *Offerings at the Wall: Artifacts from the Vietnam Veterans Memorial Collection.* Atlanta: Turner Publishing, 1995.

Allgor, Catherine. *A Perfect Union: Dolley Madison and the Creation of the American Nation.* New York: Henry Holt & Co., 2006.

Allgor, Catherine. *Parlor Politics: In Which the Ladies of Washington Help Build a City and a Government.* Charlottesville, VA: University Press of Virginia, 2000.

Alsop, Joseph. "Dining-Out Washington." In *Katharine Graham's Washington,* edited by Katharine Graham. New York: Vintage Books, 2002.

Altshuler, David, editor. *The Jews of Washington, D.C.: A Communal History Anthology.* Chappaqua, NY: Rossell Books, 1985.

"Appeal to Wilson for Negro Clerks." *New York Times,* August 18, 1913.

Arnold, R. Douglas and Nicholas Carnes. "Holding Mayors Accountable: New York's Executives from Koch to Bloomberg." *American Journal of Political Science* 56:4 (2012): 949–963.

Attoe, Wayne. "Theory, Criticism, and History of Architecture." In *Introduction to Architecture,* edited by James C. Snyder and Anthony J. Catanese. New York: McGraw-Hill, 1979.

Bacon, Edmund N. *Design of Cities.* New York: Viking Press, 1967.

Badeau, Adam. "Gen. Badeau's Letter." *New York Times,* May 15, 1887.

Badger, Emily. "The People's Transit." *Miller-McCune,* March/April, 2011.

Bai, Matt. "The Insiders." *New York Times Magazine,* June 7, 2009.

Banfield, Edward C. and James Q. Wilson. *City Politics.* Cambridge, MA: Harvard University Press and the M.I.T. Press, 1963.

Banner, James M., Jr. "The Capital and the State: Washington, D.C., and the Nature of American Government." In *A Republic for the Ages: The United States Capitol and the Political Culture of the Early Republic*, edited by Donald R. Kennon. Charlottesville, VA: University Press of Virginia, 1999.

Barber, Lucy G. *Marching on Washington: The Forging of an American Political Tradition.* Berkeley, CA: University of California Press, 2002.

Barras, Jonetta Rose. *The Last of the Black Emperors: The Hollow Comeback of Marion Barry in the New Age of Black Leaders.* Baltimore: Bancroft Press, 1998.

Bath, Georgina. "Visible Storage at the Smithsonian American Art Museum." In *The Manual of Museum Management,* edited by Gail Dexter Lord and Barry Lord. Second edition. Lanham, MD: AltaMira Press, 2009.

Bednar, Michael. *L'Enfant's Legacy: Public Open Spaces in Washington, D.C.* Baltimore: John's Hopkins University Press, 2006.

Bello, Marisol. "Obama, Laura Bush Break Ground for African American Museum." *USA Today*. February 22, 2012.

Berg, Scott W. *Grand Avenues: The Story of the French Visionary Who Designed Washington, D.C.* New York: Vintage Books, 2007.

Biegler, Caitlin. "A Big Gap: Income Inequality in the District Remains One of the Highest in the Nation." *DC Fiscal Policy Institute, Center on Budget and Policy Priorities,* March 8, 2012. Accessed December 9, 2013. http://www.dcfpi.org/wp-content/uploads/2012/03/03-08-12incomeinequality1.pdf.

Black Georgetown Remembered: A Documentary Video on the History of the Black Georgetown Community. Directed by David W. Powell. Washington, D.C.: Georgetown University, 1989.

Black, Graham. *The Engaging Museum: Developing Museums for Visitor Involvement.* New York: Routledge, 2005.

Bodnar, John. *Remaking America; Public Memory, Commemoration, and Patriotism in the Twentieth Century.* Princeton: Princeton University Press, 1992.

Borchert, James. *Alley Life in Washington: Family, Community, Religion, and Folklife in the City, 1850–1970.* Urbana, IL: University of Illinois Press, 1980.

Bouker, Jon, Brooke DeRenzis, Julia Friedman, David F. Garrison, Alice M. Rivlin, and Garry Young. "Building the Best Capital City in the World." Research Report. Washington D.C.: Brookings Institution, 2008. Accessed December 9, 2013. http://www.brookings.edu/research/reports/2008/12/18-dc-revitalization-garrison-rivlin.

Bowling, Kenneth R. *The Creation of Washington, D.C.: The Idea and Location of the American Capital.* Fairfax, VA: George Mason University Press, 1991.

Bowling, Kenneth R. "A Capital before a Capitol: Republican Visions." In *A Republic for the Ages: The United States Capitol and the Political Culture of the Early Republic*, edited by Donald R. Kennon. Charlottesville, VA: University Press of Virginia, 1999.

Bowling, Kenneth R. "'The Year 1800 Will Soon Be upon Us': George Washington and the Capitol." In *A Republic for the Ages: The United States Capitol and the Political Culture of the Early Republic*, edited by Donald R. Kennon. Charlottesville, VA: University Press of Virginia, 1999.

Bowling, Kenneth R. "From 'Federal Town' to 'National Capital': Ulysses S. Grant and the Reconstruction of Washington, D.C." *Washington History* 14:1 (Spring/ Summer 2002): 8–25.

Branch, Taylor. *Parting the Waters: America in the King Years, 1954–1963.* New York: Simon and Shuster, 1989.

Brown, DeNeen. "The 'New' Washington Dinner Party." *Washington Post Magazine,* June 10, 2012.

Brown, Karen Destorel. "Expanding Affordable Housing through Inclusionary Zoning: Lessons from the Washington Metropolitan Area." Center on Urban and Metropolitan Policy. Washington, D.C.: Brookings Institution, 2001. http://www.brookings.edu/~/media/research/files/reports/2001/10/metropolitanpolicy%20brown/inclusionary.pdf.

Brown, Letitia Woods. *Free Negroes in the District of Columbia, 1790–1846.* New York: Oxford University Press, 1972.

Browne, Charles F. M. *A Short History of the British Embassy at Washington, D.C., U.S.A.* Washington, D.C.: Gibson Bros., 1930.

Brownell, Charles E. "Thomas Jefferson's Architectural Models and the United States Capitol." In *A Republic for the Ages: The United States Capitol and the Political Culture of the Early Republic,* edited by Donald R. Kennon. Charlottesville, VA: University Press of Virginia, 1999.

Bruske, Ed. "Shepherd Park: Activism Working within Tradition." *Washington Post,* August 15, 1987.

Burleigh, Nina. *The Stranger and the Statesman: James Smithson, John Quincy Adams, and the Making of America's Greatest Museum: The Smithsonian.* New York: HarperCollins, 2004.

Busey, Samuel C. *Personal Reminiscences and Recollections of Forty-Six Years' Membership in the Medical Society of the District of Columbia, and Residence in This City, with Biographical Sketches of Many of the Deceased Members.* Philadelphia: Dornan, 1895.

Bushong, William B. "Glenn Brown and the Planning of the Rock Creek Valley." *Washington History* 14:1 (2002): 56–71.

"Cab Meters Rule Upheld in Reply to House Attack." *Washington Post,* January 7, 1932.

Cadaval, Olivia. "The Latino Community: Creating an Identity in the Nation's Capital." In *Urban Odyssey: A Multicultural History of Washington, D.C.,* edited by Francine Curro Cary. Washington, D.C.: Smithsonian Institution Press, 1996.

Callahan, Christopher. "Lagging Behind." *American Journalism Review,* August/ September 2004. Accessed August 23, 2011. http://www.ajr.org/article.asp?id=3738.

Caplan, Marvin. "Eat Anywhere!" *Washington History* 1:1 (Spring 1989): 24–39.

Caro, Robert. *The Power Broker: Robert Moses and the Fall of New York.* New York: Alfred A. Knopf, 1974.

Carpenter, Frank G. *Carp's Washington.* New York: McGraw-Hill, 1960.

Center on Urban and Metropolitan Policy. "A Region Divided: The State of Growth in Greater Washington, D.C." Research Report. Washington, D.C.: Brookings Institution, 1999.

Ceruzzi, Paul. *Internet Alley: High Technology in Tysons Corner, 1945–2005.* Cambridge: MIT Press, 2008.

Chacko, Elizabeth. "Washington D.C.: From Biracial City to Multiethnic Gateway." In *Migrants to the Metropolis: The Rise of Immigrant Gateway Cities*, edited by Marie Price and Lisa Benton-Short. Syracuse, NY: Syracuse University Press, 2008.

Cherkasky, Mara. "'For Sale to Colored': Racial Change on S Street, N.W." *Washington History* 8:2 (Fall/Winter 1996): 40–57.

Childs, Arcynta Ali. "Tour the American History Museum with an American Girl." *Smithsonian.com*, March 1, 2011. Accessed December 9, 2013. http://blogs.smithsonianmag.com/aroundthemall/2011/03/tour-the-american-history-museum-with-an-american-girl/.

Chow, Esther Ngan-ling. "From Pennsylvania Avenue to H Street, NW: The Transformation of Washington's Chinatown." In *Urban Odyssey: A Multicultural History of Washington, D.C.,* edited by Francine Curro Cary. Washington, D.C.: Smithsonian Institution Press, 1996.

Citizens Financial Report. Office of Chief Financial Officer, Government of the District of Columbia, 2011.

"Civil Rights March." NBC News Archives, August 28, 1963. Accessed December 7, 2013. http://www.nbcuniversalarchives.com/nbcuni/clip/5112485941_s01.do.

Clapper, Olive Ewing. *Washington Tapestry.* New York: McGraw-Hill, 1946.

Clark-Lewis, Elizabeth. "'For a Real Better Life': Voices of African American Women Migrants, 1900–1930." In *Urban Odyssey: A Multicultural History of Washington, D.C.,* edited by Francine Curro Cary. Washington, D.C.: Smithsonian Institution Press, 1996.

Clement, Bell. "Pushback: The White Community's Dissent from 'Bolling.'" *Washington History* 16:2 (Fall/Winter 2004): 86–109.

"Climax of Folly." *Washington Post,* May 2, 1894.

Cobb, Amanda J. "The National Museum of the American Indian: Sharing the Gift." *American Indian Quarterly* 29:3/4 (Summer/Fall 2005): 361–383.

Cohen, Andrew. "There's America—and Then There's Washington." *The Atlantic.* October 27, 2011. Accessed October 28, 2011. http://www.theatlantic.com/politics/archive/2011/10/theres-america-and-then-theres-washington/247442/.

Cohen, Patricia. "Making Way for a Dream in the Nation's Capital." *New York Times,* February 22, 2012.

"Congressional Doings Told Briefly." *Washington Post,* January 7, 1915.

Congressional Management Foundation. *House Staff Employment Study.* 2002. Accessed August 22, 2011. http://assets.sunlightfoundation.com.s3.amazonaws.com/policy/staff%20salary/2002%20House%20Staff%20employment%20Study.pdf.

Congressional Quarterly Almanac. Washington, D.C.: Congressional Quarterly Press, various dates.

Connors, Jill. Editor. *Growing Up in Washington, D.C.: An Oral History.* Charleston, SC: Arcadia Publishing, 2001.

Copeland, Libby. "Shadow Delegation Toils in Obscurity for D.C.'s Day in the Sun." *Washington Post,* January 16, 2007.

"Coxey Silenced by Police." *New York Times,* May 2, 1894.

"Coxey's Army Starts." *Washington Post,* March 26, 1894.

Craig, Tim. "Gridlock? On D.C. issues, Gray and Issa Defy Spirit of the Times." *Washington Post,* April 22, 2012.

Crew, Spencer R. "Melding the Old and the New: The Modern African American Community, 1930–1960." In *Urban Odyssey: A Multicultural History of Washington, D.C.,* edited by Francine Curro Cary. Washington, D.C.: Smithsonian Institution Press, 1996.

Crothers, A. Glenn. "The 1846 Retrocession of Alexandria: Protecting Slavery and the Slave Trade in the District of Columbia." In *In the Shadow of Freedom: The Politics of Slavery in the National Capital,* edited by Paul Finkelman and Donald R. Kennon. Athens: Ohio University Press, 2011.

Dahl, Robert A. "The Myth of the Presidential Mandate." *Political Science Quarterly* 105:3 (1990): 355–372.

Danver, Steven L. *Revolts, Protests, Demonstrations, and Rebellions in American History: An Encyclopedia.* [Electronic resource]. Santa Barbara, CA: ABC-CLIO LLC, 2011.

Davidson, Lee. "Chaffetz Giving 'Cot-Side' Chats." *Deseret News,* February 27, 2009.

Dawson, Michael C. *Behind the Mule: Race and Class in African-American Politics.* Princeton, NJ: Princeton University Press, 1994.

"DC vs. VA vs. MD." *Washingtonian Magazine,* October, 2012.

DC Mythbusting. "DC Mythbusting: International City." January 19, 2010. Accessed October 2, 2012. http://www.welovedc.com/2010/01/19/dc-mythbusting-international-city/.

Dear, Michael and Nicholas Dahmann. "Urban Politics and the Los Angeles School of Urbanism." In *The City, Revisited: Urban Theory from Chicago, Los Angeles, and New York,* edited by Dennis R. Judd and Dick Simpson. Minneapolis: University of Minnesota Press, 2011.

Dearborn, Philip M. and Stephanie Richardson. "Home Buyer Credit Widely Used." Research Report, *Greater Washington Research Center,* May 5, 1999.

DeBonis, Mike. "D.C. Keeps Wary Eye on House GOP." *Washington Post,* January 28, 2011.

DeBonis, Mike and Ben Pershing. "Vote Didn't End Budget-Autonomy Debate." *Washington Post,* April 28, 2013.

Deford, Frank. "Home without Homeowners: Washington, D.C." *Sports Illustrated,* July 2, 1979.

"Delighted with New-York." *New York Times,* August 25, 1893.

DePillis, Lydia. "More Disneyfication Coming to Chinatown." *Washington City Paper,* November 15, 2011. Accessed December 29, 2012. http://www.washingtoncitypaper.com/blogs/housingcomplex/2011/11/15/more-disneyfication-coming-to-chinatown/.

Derthick, Martha. *City Politics in Washington, D.C.* Joint Center for Urban Studies of the Massachusetts Institute of Technology and Harvard University. Cambridge: Harvard University Press, 1962.

Designing the Nation's Capital: The 1901 Plan for Washington, D.C. Washington, D.C.: U.S. Commission on Fine Arts, 2006.

Dickens, Charles. *American Notes for General Circulation.* New York: St. Martin's Press, 1985 [1842].

Diner, Hasia R. and Steven J. Diner. "Washington's Jewish Community: Separate but Not Apart." In *Urban Odyssey: A Multicultural History of Washington, D.C.*, edited by Francine Curro Cary. Washington, D.C.: Smithsonian Institution Press, 1996.

Diner, Steven J. "From Jim Crow to Home Rule." *The Washington Quarterly* 13:1 (New Year's 1989): 90–101.

Diner, Steven J. "The City under the Hill." *Washington History* 8:1 (Spring/Summer 1996): 54–61.

Dingle, Mona E. "*Gemeinschaft und Gemutlichkeit*: German American Community and Culture, 1850–1920." In *Urban Odyssey: A Multicultural History of Washington, D.C.*, edited by Francine Curro Cary. Washington, D.C.: Smithsonian Institution Press, 1996.

Dinh, Viet D. and Adam H. Charnes. "The Authority of Congress to Enact Legislation to Provide the District of Columbia with Voting Representation in the House of Representatives." Testimony submitted to the Committee on Government Reform, U.S. House of Representatives. 2004. Accessed November 22, 2011. http://www.dcvote.org/trellis/research/finaldinhopinion.cfm.

District of Columbia. *Citizens Financial Report*. Office of Chief Financial Officer. Washington, D.C., 2011.

District of Columbia. *Comprehensive Annual Financial Report for Fiscal Year 2011.* Office of the Chief Financial Officer. Washington, D.C., 2010.

Doss, Erika. *Memorial Mania; Public Feeling in America.* Chicago: University of Chicago Press, 2010.

Douglas, Danielle. "Can the Private Sector Save Washington?" *Washington Post Capital Business Section*, December 19, 2011.

Douglas, Danielle. "Uneven Upswing for D.C. Area Tourism." *Washington Post,* May 14, 2012.

Douthat, Ross. "Washington versus America." *New York Times,* September 25, 2012.

Dreier, Peter, John H. Mollenkopf, and Todd Swanstrom. *Place Matters: Metropolitics for the Twenty-First Century.* Lawrence, KS: University of Kansas Press, 2005.

Du Lac, J. Freedom. "A Stirring Moment in Jazz History to Echo in Turkish Embassy." *Washington Post,* February 4, 2011.

Dumain, Emma. "Commuter Tax Is Panned." *Roll Call,* July 25, 2012.

Dvorak, Petula. "Vietnam Wall Visitor Center Approved." *Washington Post,* August 4, 2006.

Earman, Cynthia D. "Remembering the Ladies: Women, Etiquette, and Diversions in Washington City, 1800–1814." *Washington History* 12:1 (Spring/Summer 2000): 102–117.

Eberle, Josef. "A German Editor's Impressions of 'The City without a Mayor.'" *Washington Post,* June 5, 1949.

Edgerly, Louisa, Amoshaun Toft, and Mary Lynn Veden. "Social Movements, Political Goals, and the May 1 Marches: Communicating Protest in Polysemous Media Environments." *International Journal of Press/Politics* 16:3 (2011): 314–334.

Edwards, George C., III. *On Deaf Ears: The Limits of the Bully Pulpit.* New Haven: Yale University Press, 2003.

Eisen, Jack and LaBarbara Bowman. "House Joins Senate to Kill Chancery Bill." *Washington Post,* December 21, 1979.

El Nasser, Haya. "Young and Educated Show Preference for Urban Living." *USA Today*, April 1, 2011.

Elaigwu, J. Isawa. "Abuja, Nigeria." In *Finance and Governance of Capital Cities in Federal Systems,* edited by Enid Slack and Rupak Chattopadhyay. Montreal: McGill-Queen's University Press, 2009.

Elliott, Justin and Zachary Roth. "Shadow Congress: More than 170 Former Lawmakers Ply the Corridors of Power as Lobbyists." *Talking Points Memo,* June 1, 2010. Accessed December 3, 2013. http://tpmmuckraker.talkingpointsmemo.com/2010/06/shadow_congress_former_lawmakers_become_lobbyists.php.

Eshleman, J. Ross, Barbara G. Cashion, and Laurence A. Basirico. *Sociology: An Introduction,* fourth ed. New York, NY: HarperCollins College Publishers, 1993.

"Eustis Warmly Received." *New York Times,* May 7, 1893.

Evelyn, Douglas E. and Paul Dickson. *On This Spot: Pinpointing the Past in Washington, D.C.* Washington, D.C.: Farragut Publishing Company, 1992.

Evenson, Norma. "Monumental Spaces." In *The Mall in Washington, 1791–1991,* edited by Richard Longstreth. Washington, D.C.: National Gallery of Art, 1991.

Ewing, Heather. *The Lost World of James Smithson: Science, Revolution, and the Birth of the Smithsonian.* New York: Bloomsbury, 2007.

Ewing, Heather and Amy Ballard. *A Guide to Smithsonian Architecture.* Washington, D.C.: Smithsonian Books, 2009.

Fahrenthold, David A. "The House Is Their Home." *Washington Post,* February 15, 2011.

Farhi, Paul. "Catering to the C-suite." *Washington Post,* April 26, 2013.

Farrar, Margaret E. *Building the Body Politic: Power and Urban Space in Washington, D.C.* Urbana: University of Illinois Press, 2008.

Fauntroy, Michael K. *Home Rule or House Rule? Congress and the Erosion of Local Governance in the District of Columbia.* Dallas, TX: University Press of America, 2003.

Feiss, Carl. "Washington, D.C.: Symbol and City." In *World Capitals: Toward Guided Urbanization*, edited by Hanford Wentworth Eldredge. New York: Anchor Press, 1975.

Fenno, Richard F., Jr. *Home Style: House Members in Their Districts.* New York: HarperCollins Publishers, 1978.

Field, Cynthia R. and Jeffrey T. Tilman. "Creating a Model for the National Mall: The Design of the National Museum of Natural History." *Journal of the Society of Architectural Historians* 63:1 (2004): 52–73.

Fiorina, Morris. *Congress: Keystone of the Washington Establishment.* Revised edition. New Haven: Yale University Press, 1989.

Fisher, Marc. "Chevy Chase, 1916: For Everyman, a New Lot in Life." *Washington Post,* February 15, 1999.

Fisher, Marc. "District No Joke Now." *Washington Post,* October 26, 2011.

Fisher, Marc. "Does Culture Follow the Census?" *Washington Post,* April 11, 2011.

Fitzpatrick, Sandra and Maria R. Goodwin. *The Guide to Black Washington: Places and Events of Historical and Cultural Significance in the Nation's Capital.* Rev. edition. New York: Hippocrene Books, 2001.

Fletcher, Kenneth R. "A Brief History of Pierre L'Enfant and Washington, D.C." *Smithsonian Magazine*, May 1, 2008. Accessed December 9, 2013. http://www.smithsonianmag.com/arts-culture/brief-history-of-lenfant.html.

Florida, Richard. "The 25 Most Economically Powerful Cities in the World." *The Atlantic*, September 15, 2011. Accessed December 8, 2013. http://www.theatlanticcities.com/jobs-and-economy/2011/09/25-most-economically-powerful-cities-world/109/#slide9.

Florida, Richard. *The Rise of the Creative Class: And How It's Transforming Work, Leisure, Community and Everyday Life.* New York: Basic Books, 2002.

Florida, Richard. "Where the Creative Class Jobs Will Be." *The Atlantic.com*, August 25, 2010. Accessed December 8, 2013. http://www.theatlantic.com/business/archive/2010/08/where-the-creative-class-jobs-will-be/61468//.

Florida, Richard. "Where the Jobs Will Be." *The Atlantic.com,* August 17, 2010. Accessed December 7, 2013. http://www.theatlantic.com/business/archive/2010/08/where-the-jobs-will-be/61459/.

Fogelson, Robert M. *Downtown: Its Rise and Fall, 1880–1950.* New Haven: Yale University Press, 2003.

Fowler, Glenn. "David Lloyd Kreeger Dead at 81; Insurance Official and Arts Patron." *New York Times,* November 20, 1990.

Franco, Barbara. "The Challenge of a City Museum for Washington, D.C." *Washington History* 15:1 (Spring/Summer 2003): 4–25.

Frank, Thomas. "The Bleakness Stakes." *Harper's Magazine,* November 2011.

Franklin, Jay. "Main Street-on-Potomac." In *Katharine Graham's Washington*, edited by Katharine Graham. New York: Vintage Books, 2002.

Freeman, Tyson. "The 1980s: (Too) Easy Money Fuels a Building Boom." *National Real Estate Investor,* September 30, 1999. Accessed December 9, 2013. http://nreionline.com/mag/real_estate_easy_money_fuels/.

Frey, William H. *Brookings Institution and University of Michigan Social Science Data Analysis Network's Analysis of 2005–9 American Community Survey and 2000 Census Decennial Census Tract Data.* Accessed November 15, 2012. http://www.psc.isr.umich.edu/dis/census/segregation.html.

Frey, William H. "Millennial and Senior Migrants Follow Different Post-Recession Paths." Brookings Institution Report, November 15, 2013. Accessed November 20, 2013. http://www.brookings.edu/research/opinions/2013/11/15-millennial-senior-post-recession-frey.

Friedman, Emily. "Mount Pleasant Riots: May 5 Woven into Neighborhood's History." WAMU, May 5, 2011. Accessed January 4, 2013. http://wamu.org/news/11/05/05/mount_pleasant_riots_may_5_woven_into_neighborhoods_history.php.

Friedman, Samantha, Audrey Singer, Marie Price, and Ivan Cheung. "Race, Immigrants, and Residence: A New Racial Geography of Washington, D.C." *The Geographical Review* 95:2 (2005): 210–230.

Fuller, Stephen S. "Northern Virginia's Economic Transformation." November 2011. Accessed December 9, 2013. http://cra.gmu.edu/pdfs/studies_reports_presentations/By_The_Numbers_NoVa_Drives_Area_Growth.pdf.Fuller, Stephen S. "The Economic and Fiscal Impact of Foreign Missions on the Nation's Capital." National Capital Planning Commission, June 6, 2002. Accessed January 8, 2013. http://www.ncpc.gov/DocumentDepot/Publications/ForeignMissions/Foreign_Missions_Impact.pdf.

Fuller, Stephen S. "The Impact of the Consortium of Universities of the Washington Metropolitan Area on the Economies of the Washington Metropolitan Area and District of Columbia." May 2011. Accessed December 9, 2013. http://cra.gmu.edu/pdfs/researach_reports/recent_reports/Economic_Impacts_of_Washington_Consortium_Universities.pdf.

Fulwood, Sam III. "Why Blacks Support Mayor Barry." *Los Angeles Times,* August 11, 1990.

Furgurson, Ernest B. *Freedom Rising: Washington in the Civil War.* New York: Alfred A. Knopf, 2004.

Gaither, Edmund Barry. "'Hey! That's Mine': Thoughts on Pluralism and American Museums." In *Reinventing the Museum: Historical and Contemporary Perspectives on the Paradigm Shift*, edited by Gail Anderson. Walnut Creek, CA: AltaMira Press, 2004.

Gale, Dennis E. *Washington, D.C.: Inner-City Revitalization and Minority Suburbanization.* Philadelphia: Temple University Press, 1987.

Gandhi, Natwar M. "Tax Rates and Tax Burdens in the District of Columbia—A Nationwide Comparison, 2010." Office of the Chief Financial Officer, Government of the District of Columbia, 2010. http://www.cfo.dc.gov/cfo/frames.asp?doc=/cfo/lib/ cfo/10study.pdf.

Garance Franke-Ruta. "Facts and Fictions of D.C.'s Gentrification." *The Atlantic Cities,* August 10, 2012. Accessed December 9, 2013. http://www.theatlanticcities.com/politics/2012/08/facts-and-fictions-gentrification-dc/2914/.

Garnick, Darren and Ilya Mirman. "If the Walls Could Talk." *Slate,* October 28, 2010. Accessed September 9, 2011. http://www.slate.com/id/2272482/.

Garreau, Joel. *Edge City: Life on the New Frontier.* New York: Doubleday, 1991.

Gavin, Patrick. "Brokaw Says 'No Thanks' To WHCD." *Politico's Guide to the White House Correspondents' Dinner,* April 26, 2013.

Gay, Claudine. "Putting Race in Context: Identifying the Environmental Determinants of Black Racial Attitudes." *American Political Science Review* 98:4 (2004): 547–562.

Geertz, Clifford. *Local Knowledge: Further Essays in Interpretive Anthropology.* New York: Basic Books, 1983.

Gelertner, Mark. *A History of American Architecture: Buildings in Their Cultural and Technological Context.* Hanover, NH: University Press of New England, 1999.

Gendreau, Andrée. "Museums and Media: A View from Canada." *The Public Historian* 31:1 (Winter 2009): 35–45.

Gerstle, Gary. *American Crucible: Race and Nation in the Twentieth Century.* Princeton: Princeton University Press, 2001.

Gandhi, Natwar M., Yesim Yilmaz, Robert Zahradnik, and Marcy Edwards. "Washington, District of Columbia, United States of America." In *Finance and Governance of Capital Cities in Federal Systems,* edited by Enid Slack and Rupak Chattopadhyay. Montreal: McGill-Queen's University Press, 2009.

Gibson, Campbell and Kay Jung. *Historical Census Statistics on Population Totals by Race, 1790 to 1990, and by Hispanic Origin, 1970 to 1990, for Large Cities and Other Urban Places in the United States.* Working Paper No. 76. Washington, D.C.: U.S. Census, February 2005.

Gibson, Campbell and Kay Jung. *Historical Census Statistics on Population Totals by Race, 1790 to 1990, and by Hispanic Origin, 1970 to 1990, for the United States, Regions, Divisions, and States.* Working Paper No. 56. Washington, D.C.: U.S. Census, September 2002.

Gillette, Howard, Jr. *Between Justice and Beauty: Race, Planning, and the Failure of Urban Policy in Washington, D.C.* Baltimore: Johns Hopkins University Press, 1995.

Gillette, Howard, Jr. "Protest and Power in Washington, D.C.: The Troubled Legacy of Marion Barry." In *African-American Mayors: Race, Politics, and the American City,* edited by David R. Colburn and Jeffrey S. Adler. Urbana, Ill: University of Illinois Press, 2001.

Gillette, Howard, Jr. and Alan M. Kraut. "The Evolution of Washington's Italian American Community." In *Urban Odyssey: A Multicultural History of Washington, D.C.,* edited by Francine Curro Cary. Washington, D.C.: Smithsonian Institution Press, 1996.

Glazer, Nathan. "Monuments, Modernism, and the Mall." In *The National Mall: Rethinking Washington's Monumental Core,* edited by Nathan Glazer and Cynthia R. Field. Baltimore: Johns Hopkins University Press, 2008.

Goldfarb, Zachary A. 2007. "Volkswagen Moving to Herndon." *Washington Post,* September 6.

"Good-Bye to Sackville." *New York Times,* October 31, 1888.

Goode, James M. *Best Addresses: A Century of Washington's Distinguished Apartment Houses.* Washington, D.C.: Smithsonian Institution Press, 1988.

Goode, James. *Washington Sculpture; A Cultural History of Outdoor Sculpture in the Nation's Capital.* Baltimore: Johns Hopkins University Press, 2008.

"Great Seneca Science Corridor." Maryland-National Capital Park and Planning Commission. Accessed January 8, 2013. http://www.montgomeryplanning.org/community/gaithersburg/index.shtm.

Greater Washington Board of Trade. *Greater Washington 2010 Regional Report.* Washington, D.C., 2009.

Greek, Mark S. *Washington, D.C. Protests: Scenes from Home Rule to the Civil Rights Movement.* Charleston, S.C.: The History Press, 2009.

Green, Constance McLaughlin. *The Secret City: A History of Race Relations in the Nation's Capital.* Princeton: Princeton University Press, 1967.

Green, Constance McLaughlin. *Washington, Vol. 1: Village and Capital, 1800–1878.* Princeton, NJ: Princeton University Press, 1962.

Green, Constance McLaughlin. *Washington, Vol. 2: Capital City, 1879–1950.* Princeton, NJ: Princeton University Press, 1963.

Greene, Richard P. and James B. Pick. *Exploring the Urban Community: A GIS Approach*. Second Edition. Boston: Prentice Hall, 2006.

Greenfield, Meg. *Washington.* New York: Public Affairs, 2001.

Gruber, J. 2003. "Smithsonian Remakes Its Transportation Exhibit: Corporate Sponsors Help with First Major Change in Four Decades." *Trains* 63:3 (March): 76–77.

Guinther, John. *Direction of Cities.* New York: Penguin Books, 1996.

Gutheim, Frederick and Antoinette J. Lee. *Worthy of the Nation: Washington, D.C., from L'Enfant to the National Capital Planning Commission.* Second edition. Baltimore, MD: Johns Hopkins University Press, 2006.

Hackett, Beatrice Nied. "'We Must Become Part of the Larger American Family': Washington's Vietnamese, Cambodians, and Laotians." In *Urban Odyssey: A Multicultural History of Washington, D.C.*, edited by Francine Curro Cary. Washington, D.C.: Smithsonian Institution Press, 1996.

Hafertepe, Kenneth. "An Inquiry into Thomas Jefferson's Ideas of Beauty." *Journal of the Society of Architectural Historians,* 59:2 (2000): 216–231.

Hagedorn, David. "Guess Who's Coming . . ." *Washington Post,* March 16, 2011.

Halsey, Ashley, III. "We're No. 1 in Traffic Gridlock." *Washington Post,* September 27, 2011.

Halzack, Sarah. "D.C. Area Bucks Trend of 'Job Sprawl.'" *Washington Post,* April 22, 2013.

Hamilton, Alexander James Madison and John Jay. *The Federalist Papers.* New York: Signet Classics, 2003 [1788].

Harris, Ann Sutherland. *Seventeenth Century Art & Architecture.* London: Laurence King Publishing, 2005.

Harris, Charles Wesley. *Congress and the Governance of the Nation's Capital: The Conflict of Federal and Local Interests.* Washington, D.C.: Georgetown University Press, 1995.

Harris, Charles Wesley. "In Whose Interest? Congressional Funding for Washington in the Home-Rule Era." *Washington History* (Spring/Summer 1996): 62–70.

Harrison, Robert. *Washington during Civil War and Reconstruction: Race and Radicalism.* New York: Cambridge University Press, 2011.

Harrold, Stanley. *Subversives: Antislavery Community in Washington, D.C., 1828–1865.* Baton Rouge: Louisiana State University Press, 2003.

Hartz, Louis. *The Liberal Tradition in America.* New York: Harcourt, Brace, & World, 1955.

Hawkins, Don Alexander. "Masonic Symbols in the L'Enfant Plan: An Examination of Recent Publications." *Washington History* 21 (2009): 100–105.

Hayes, Christopher. "Why Washington Doesn't Care about Jobs." *The Nation,* March 3, 2011.

Heller, Allan M. *Monuments and Memorials of Washington, D.C.* Atglen, PA: Schiffer Publishing, Ltd., 2006.

Hemrick, Eugene F. *One Nation under God: Religious Symbols, Quotes, and Images in Our Nation's Capital.* Huntington, IN: Our Sunday Visitor, 2001.

Hermann, Peter. "As D.C. Changes, Police Try to Adapt." *Washington Post,* August 5, 2012.

Hesse, Monica. "Dinner at America's Table." *Washington Post*, October 13, 2011.

Heyman, I. Michael. "Museums and Marketing." *Smithsonian*, January 1998. http://www.smithsonianmag.com/history-archaeology/heyman_jan98-abstract.html.

Hicks, Josh. "How Much Did Closing the Government Cost?" *Washington Post,* October 21, 2013.

Highsmith, Carol M. and Ted Landphair. *Embassies of Washington.* Washington, D.C.: Preservation Press, 1992.

Hilton, Shani O. "Confessions of a Black Gentrifier." *Washington City Paper,* March 18, 2011.

Hines, Tomas S. "The Imperial Mall: The City Beautiful Movement and the Washington Plan of 1901–1902." In *The Mall in Washington, 1791–1991,* edited by Richard Longstreth. Washington, D.C.: National Gallery of Art, 1991.

Holland, Jesse J. *Black Men Built the Capitol: Discovering African-American History in and around Washington, D.C.* Guilford, CT: The Globe Pequot Press, 2007.

Homan, Timothy R. "Unemployment Rate in Washington's Ward 8 Is Highest in U.S." *Bloomberg News,* March 30, 2011. Accessed October 26, 2012. http://www.bloomberg.com/news/2011-03-30/unemployment-rate-in-washington-s-ward-8-is-highest-in-u-s-.html.

Hopkinson, Natalie. *Go-Go Live: The Musical Life and Death of Chocolate City.* Durham: Duke University Press, 2012.

Horton, James Oliver. "The Genesis of Washington's African American Community." In *Urban Odyssey: A Multicultural History of Washington, D.C.*, edited by Francine Curro Cary. Washington, D.C.: Smithsonian Institution Press, 1996.

Hughes, Langston. "The Big Sea." In *Literary Washington, D.C.,* edited by Patrick Allen. San Antonio: Trinity University Press, 2012.

Hyman, Louis. *Debtor Nation: The History of America in Red Ink.* Princeton: Princeton University Press, 2011.

Ignatiev, Noel. *How the Irish Became White.* New York: Routledge, 1995.

IMLS Office of Strategic Partnerships. *Museums, Libraries and 21st Century Skills.* Washington, DC: Institute of Museum and Library Services, 2009.

"Insist upon Race Law." *Washington Post,* July 17, 1913.

Jackson, Donald E. "L'Enfant's Washington: An Architect's View." *Washington History* 50 (1978): 398–420.

Jackson, Kenneth T. *Crabgrass Frontier: The Suburbanization of the United States.* New York: Oxford University Press, 1985.

Jackson, Robert L. "Smithsonian's New Ocean Planet Exhibit Catches a Wave: Corporate Sponsorship." *The Baltimore Sun*. June 7, 1995.

Jacob, Kathryn Allamong. *Capital Elites: High Society in Washington, D.C., after the Civil War.* Washington: Smithsonian Institution Press, 1995.

Jacob, Kathryn Allamong. "'Like Moths to a Candle': The Nouveaux Riches Flock to Washington, 1870–1900." In *Urban Odyssey: A Multicultural History of Washington, D.C.*, edited by Francine Curro Cary. Washington, D.C.: Smithsonian Institution Press, 1996.

Jacobs, Jane. *The Death and Life of Great American Cities.* New York: Random House, 1961.

Jacobson, Matthew Frye. *Whiteness of a Different Color: European Immigrants and the Alchemy of Race.* Cambridge: Harvard University Press, 1999.

Jaffe, Harry S. and Tom Sherwood. *Dream City: Race, Power, and the Decline of Washington, D.C.* New York: Simon & Schuster, 1994.

James, Portia. "Building a Community-Based Identity at Anacostia Museum." In *Heritage, Museums and Galleries: An Introductory Reader*, edited by Gerard Corsane. New York: Routledge, 2005.

Janken, Kenneth R. "Rayford Logan: The Golden Years." *Negro History Bulletin*, 61:3/4 (1998).

Janofsky, Michael. "The 1994 Campaign: The Comeback Man in the News.'" *New York Times,* September 14, 1994.

Jenkins, Jeffrey A. and Charles Stewart, III. *Fighting for the Speakership: The House and the Rise of Party Government.* Princeton: Princeton University Press, 2013.

Jenkins, Kent, Jr. "D.C.'s Clout Dwindling on the Hill, Lawmakers Warn." *Washington Post,* July 5, 1993.

Jennings, J. L. Sibley, Jr. "Artistry as Design, L'Enfant's Extraordinary City." *The Quarterly Journal of the Library of Congress* 1979: 225–278.

Johnston, Allan. *Surviving Freedom: The Black Community of Washington, D.C., 1860–1880.* New York: Garland Publishing, Inc., 1993.

Judd, Dennis R. and Todd Swanstrom. *City Politics: The Political Economy of Urban America,* fifth edition. New York: Pearson/Longman, 2006.

Judd, Dennis R. and Todd Swanstrom. *City Politics: The Political Economy of Urban America*, eighth edition. Boston: Longman, 2012.

Kaiser, Robert G. "Big Money Created a New Capital City." *Washington Post,* April 8, 2007.

Katz, Jon. "2 Students Say Albert Was 'Drunk.'" *Washington Post,* September 13, 1972.

Keating, Joshua E. "Can You Get away with Any Crime If You Have Diplomatic Immunity?" *Foreign Policy*, February 15, 2011. Accessed October 2, 2012. http://www.foreignpolicy.com/articles/2011/02/15/can_you_get_away_with_any_crime_if_you_have_diplomatic_immunity.

Kelly, John. "Memories from the Front Lines of the Segregation Battle in the District." *Washington Post,* October 11, 2011.

Kelly, John. "Old D.C.'s Manifest Destiny: Staying Put." *Washington Post,* August 15, 2010.

Kernell, Samuel. *Going Public: New Strategies of Presidential Leadership,* fourth edition. Washington, D.C.: CQ Press, 2006.

Kerr, Audrey Elisa. *The Paper Bag Principle: Class, Colorism, and Rumor and the Case of Black Washington, D.C.* Knoxville: University of Tennessee Press, 2006.

Kinder, Donald and Nicholas Winter. "Exploring the Racial Divide: Blacks, Whites, and Opinion on National Policy." *American Journal of Political Science* 45 (April, 2011): 439–456.

King, Anthony. "The Vulnerable American Politician." *British Journal of Political Science* 27 (1997): 1–22.

Klaus, Susan L. *Links in the Chain: Greenbelt, Maryland, and the New Town Movement in America.* Washington, D.C.: George Washington University Press, 1987.

Klein, Julia M. "Two New History Museums Put Their Ideals on Display." *Chronicle of Higher Education* 51:6 (October 1, 2004): B15–16.

Knepper, Cathy D. *Greenbelt, Maryland: A Living Legacy of the New Deal.* Baltimore: Johns Hopkins University Press, 2001.

Knight, Louise W. "Changing My Mind: An Encounter with Jane Addams." *Affilia: Journal of Women & Social Work,* 21:1 (2006): 97–102.

Knorr, Jeremy. "Political Parameters: Finding a Route for the Capital Beltway, 1950–1964." *Washington History* 19/20 (2007/2008): 4–29.

Kofie, Nelson F. *Race, Class, and the Struggle for Neighborhood in Washington, D.C.* New York: Garland Publishing, 1999.

Kostof, Spiro. *The City Assembled: The Elements of Urban Form through History.* London: Thames & Hudson, 1992.

Kostof, Spiro. *The City Shaped: Urban Patterns and Meanings through History.* Boston: Bulfinch Press, 1993.

Kruft, Hanno-Walter. *A History of Architectural Theory.* New York: Princeton Architectural Press, 1994.

Kurtz, Howie and Michael Isikoff. "Congress Still Rules the Roost in District." *Washington Post,* October 25, 1981.

Lacy, Karyn R. *Blue-Chip Black: Race, Class, and Status in the New Black Middle Class.* Berkeley: University of California Press, 2007.

"Lapel Pins to Identify Congressmen." *Youngstown Vindicator,* May 19, 1975.

Larson, Stephanie Greco. *Media & Minorities: The Politics of Race in News and Entertainment.* Lanham, MD: Rowman & Littlefield Publishers, 2006.

Lee, Trymaine. "As Black Population Declines in Washington, D.C., Little Ethiopia Thrives." *Huffington Post,* April 8, 2011. Accessed November 18, 2012. http://www.huffingtonpost.com/2011/04/08/black-population-declies-dc-little-ethiopia-thrives_n_846817.html.

Leech, Margaret. *Reveille in Washington: 1860–1865.* New York: London, Harper & Bros, 1942.

Leibovich, Mark. *This Town: Two Parties and a Funeral—Plus Plenty of Valet Parking!—In America's Gilded Capital.* New York: Blue Rider Press, 2013.

Lemann, Nicholas. *The Promised Land: The Great Migration and How It Changed America.* New York: Vintage Books, 1991.

Lesko, Kathleen M., Valerie Babb, and Samuel Harvey. *Black Georgetown Remembered: A History of Its Black Community from the Founding of 'The Town of George' in 1751 to the Present Day.* Washington, D.C.: Georgetown University Press, 1991.

Lessoff, Alan. *The Nation and Its City: Politics, 'Corruption,' and Progress in Washington, D.C., 1861–1902.* Baltimore, MD: Johns Hopkins University Press, 1994.

Lethbridge, Francis D. "The Architecture of Washington, D.C." In *The AIA Guide to the Architecture of Washington, D.C.*, edited by G. Martin Moeller, Jr. Baltimore, MD: Johns Hopkins University Press, 2006.

Levey, Bob and Jane Freundel Levey. "End of the Roads." *Washington Post.* November 26, 2000.

Lewis, Catherine M. *The Changing Face of Public History: The Chicago Historical Society and the Transformation of an American Museum.* DeKalb: Northern Illinois University Press, 2005.

Lewis, David E. 2008. *The Politics of Presidential Appointments.* Princeton, NJ: Princeton University Press.

Lewis, Michael. "The Idea of the American Mall." In *The National Mall: Rethinking Washington's Monumental Core*, edited by Nathan Glazer and Cynthia R. Field. Baltimore: Johns Hopkins University Press, 2008.

Li, Wei. 1998. "Anatomy of a New Ethnic Settlement: The Chinese Ethnoburb in Los Angeles." *Urban Studies* 35(3): 479–501.

Library of Congress. 2006. "Meet Me at the Willard: Famed Hotel Is Subject of Library Display." *Information Bulletin* 65:4.

Liebow, Elliot. 2003 [1967]. *Tally's Corner: A Study of Negro Streetcorner Men.* Lanham, MD: Rowman & Littlefield Publishers.

Linenthal, Edward T. 1995. *Preserving Memory: The Struggle to Create America's Holocaust Museum.* New York: Viking Penguin.

Linenthal, Edward T. and Tom Engelhardt, editors. 1996. *History Wars: The* Enola Gay *and Other Battles for the American Past.* New York: Henry Holt and Company.

Linton, Ralph. 1936. *The Study of Man: An Introduction.* Appleton-Century-Crofts.

Longstreth, Richard. "Introduction: Change and Continuity on the Mall." In *The Mall in Washington, 1791–1991,* edited by Richard Longstreth. Washington, D.C.: National Gallery of Art, 1991.

Lopez, Mark Hugo and Daniel Dockterman. "A Growing and Diverse Population: Latinos in the Washington, DC Metropolitan Area." In *Hispanic Migration and Urban Development: Studies from Washington DC,* edited by Enrique S. Pumar. Bingley, UK: Emerald Press, 2012.

"Lord Sackville's Circular." *Washington Post,* October 10, 1895.

Lornell, Kip and Charles C. Stephenson. *The Beat! Go-Go from Washington, D.C.* Jackson, MS: University Press of Mississippi, 2009.

Lott, Trent. *Herding Cats: A Life in Politics.* New York: HarperCollins, 2005.

Lovell, Mary S. *Cast No Shadow: The Life of the American Spy Who Changed the Course of World War II.* New York: Pantheon Books, 1992.

Lowrey, Annie. "Washington's Economic Boom, Financed by You." *New York Times,* January 10, 2013.

Lubar, Steven and Kathleen M. Kendrick. *Legacies: Collecting America's History at the Smithsonian.* Washington, D.C.: Smithsonian Institution Press, 2001.

Lucy, William H. and David L. Phillips. *Tomorrow's Cities, Tomorrow's Suburbs.* Washington, D.C.: Planners Press, 2006.

Lusane, Clarence. *The Black History of the White House.* San Francisco: City Lights Books, 2011.

Lyons, Terrence. 2009. "The Ethiopian Diaspora and Homeland Conflict." *Institute for Conflict Analysis and Resolution,* George Mason University, http://portal.svt.ntnu.no/sites/ices16/Proceedings/Volume%202/Terrence%20Lyons-%20The%20Ethiopian%20Diaspora%20and%20Homeland%20Conflict.pdf.

Mabe, Matt. "The World's Most Global Cities." *Businessweek,* October 29, 2008. Accessed February 2, 2013. http://images.businessweek.com/ss/08/10/1028_global_cities/3.htm.

MacCleery, Rachel and Jonathan Tarr. "NoMa: The Neighborhood That Transit Built." *Urbanland*, February 29, 2012. Accessed December 9, 2013. http://urbanland.uli.org/Articles/2012/Jan/MacCleeryNOMA.

MacDonald, J. Fred. *Blacks and White TV: Afro-Americans in Television since 1948.* Chicago, IL: Nelson-Hall Publishers, 1983.

Mackintosh, Barry. *Rock Creek Park: An Administrative History.* Washington, D.C.: National Park Service, 1985.

McAleer, Margaret H. "'The Green Streets of Washington': The Experience of Irish Mechanics in Antebellum Washington." In *Urban Odyssey: A Multicultural History of Washington, D.C.*, edited by Francine Curro Cary. Washington, D.C.: Smithsonian Institution Press, 1996.

McCann, Joseph T. "Spillover Effect: The Assassination of Orlando Letelier." In *Terrorism on American Soil: A Concise History of Plots and Perpetrators from the Famous to the Forgotten*, edited by Joseph T. McCann. Boulder, CO: Sentient Publications, 2006.

McCartney, Robert. "A Warning to District's Political Old Guard." *Washington Post,* May 2, 2013.

McDaniel, George W. and John N. Pearce, eds. *Images of Brookland: The History and Architecture of a Washington Suburb.* Revised and enlarged by Martin Aurand. Washington, D.C.: Center for Washington Area Studies, George Washington University, 1982.

McDonald, Forrest. *Novus Ordo Seclorum.* Lawrence, KS: University Press of Kansas, 1985.

McGovern, Stephen J. *The Politics of Downtown Development: Dynamic Political Cultures in San Francisco and Washington, D.C.* Lexington: University Press of Kentucky, 1998.

McGregor, James H. S. *Washington from the Ground Up*. Cambridge, MA: Harvard University Press, 2007.

McMahon, Kevin J. *Nixon's Court: His Challenge to Judicial Liberalism and Its Political Consequences.* Chicago: University of Chicago Press, 2011.

"Made An Ambassador." *New York Times,* March 25, 1893.

Masur, Kate. *An Example for All the Land: Emancipation and the Struggle over Equality in Washington, D.C.* Chapel Hill: University of North Carolina Press, 2010.

May, Clifford D. "Washington Talk: Home Rule." *New York Times,* January 11, 1989.

May, Ernest R. *Imperial Democracy: The Emergence of America as a Great Power.* New York: Harcourt, Brace & World, Inc., 1961.

Mayhew, David. *Congress: The Electoral Connection.* New Haven: Yale University Press, 1974.

Mencken, H. L. *On Politics.* Edited by Malcolm Moos. Baltimore: Johns Hopkins University Press, 1956.

Mesta, Perle. "Bigwigs, Littlewigs, and No Wigs at All." Excerpted from *Katharine Graham's Washington*, edited by Katharine Graham. New York: Vintage Books, 2002.

Meyer, Jeffrey F. *Myths in Stone: Religious Dimensions of Washington, D.C.* Berkeley, CA: University of California Press, 2001.

Meyers, Edward M. *Public Opinion and the Political Future of the Nation's Capital.* Washington, D.C.: Georgetown University Press, 1996.

Milbank, Dana. "Journalists Gone Wild." *Washington Post,* May 1, 2011.

Milius, Peter. "Why Coates? Image . . . Unity." *Washington Post,* January 29, 1969.

Miller, Hope Ridings. *Embassy Row: The Life & Times of Diplomatic Washington.* New York: Holt, Rinehart and Winston, 1969.

Miller, Iris. *Washington in Maps: 1606–2000.* New York: Rizzoli International Publications, 2002.

Miller, Zane L. *The New Deal in the Suburbs: A History of the Greenbelt Town Program, 1935–1954.* Columbus: Ohio State University Press, 1971.

Mills, Nicolaus. *Their Last Battle: The Fight for the National World War II Memorial.* New York: Basic Books, 2004.

Mondell, William H. "Capital in 'Parliamentary Peonage,' Proponent of Vote Says." *Washington Post,* April 30, 1938.

Montes, Sue Anne Pressley. "Report Adds to Debate over Putting Meters in D.C. Cabs." *Washington Post,* July 28, 2007.

Montgomery County Main Street Heritage Trail Project. Accessed October 25, 2012. http://www.montgomerycountymd.gov/content/council/mem/ervin_v/pdfs/ervin-silverspringhistorytrail41210.pdf.

Moore, Jacqueline M. *Leading the Race: The Transformation of the Black Elite in the Nation's Capital, 1880–1920.* Charlottesville, VA: University Press of Virginia, 1999.

Moore, Thomas. *The Poetical Works of Thomas Moore with Life.* Edinburgh: Gall and Inglis, 1859.

Morello, Carol and Dan Keating. "Blacks' Majority Status Slips Away." *Washington Post,* March 25, 2011.

Morello, Carol and Dan Keating. "The New American Neighborhood." *Washington Post,* October 30, 2011.

Morello, Carol, Dan Keating, and Steve Hendrix. "Capital Hip: D.C. Is Getting Younger." *Washington Post,* May 5, 2011.

Morgenthau, Hans. *Politics among Nations,* fourth edition. New York: Knopf, 1967.

Morin, Richard and Michael Abramowitz. "Dixon Victory Points to Economic Rift between Voters." *Washington Post,* September 16, 1990.

Morley, Jefferson. *Snow-Storm in August: Washington City, Francis Scott Key, and the Forgotten Race Riot of 1835.* New York: Doubleday, 2012.

Mount Pleasant Main Street Organization. Accessed October 25, 2012. http://www.mtpleasantdc.org.

Muller, John. *Frederick Douglass in Washington, D.C.: The Lion of Anacostia.* Charleston, SC: History Press, 2012.

Mumford, Lewis. *The City in History.* San Diego: Harcourt, 1961.

National Capital Planning Commission. "Extending the Legacy: Planning America's Capital for the 21st Century." 1997. http://www.ncpc.gov/ncpc/Main(T2)/Planning(Tr2)/ExtendingtheLegacy.html.

National Capital Planning Commission. "Memorials and Museums Master Plan." 2006 [2001]. http://www.ncpc.gov/ncpc/Main(T2)/Planning(Tr2)/2MPlan.html.

National Capital Planning Commission. "Monumental Core Framework Plan." 2009. http://www.ncpc.gov/ncpc/Main(T2)/Planning(Tr2)/FrameworkPlan.html.

"National Capital Topics." *New York Times,* November 6, 1898.

National Park Service. "The Greater U Street Historic District." Accessed November 20, 2012. http://www.nps.gov/nr/travel/wash/dc63.htm.

National Park Service National Register of Historic Places. "Mount Pleasant Historic District." Accessed October 25, 2012. http://www.cr.nps.gov/nr/travel/wash/dc96.htm.

Natividad, Ivan V. "A Haven for the Homeless." *Roll Call,* December 14, 2011.

"Negro May Head D.C. School Board." *New York Times*, November 28, 1968.

Neibauer, Michael. "Howard University No Longer D.C.'s Top Employer." *Washington Business Journal,* February 4, 2011. Accessed November 2, 2012. http://www.bizjournals.com/washington/blog/2011/02/howard-no-longer-dcs-top-employer.html.

Neustadt, Richard. *Presidential Power and the Modern Presidents: The Politics of Leadership from Roosevelt to Reagan.* New York: Free Press, 1990.

Newhauser, Daniel. "Gray Arrest Highlights D.C. Battle." *Roll Call,* April 12, 2011.

Newhouse, John. "Diplomacy Inc." *Foreign Affairs* (May/June 2009): 73–92.

Newsome, Oramenta and Michael Rubinger. "The H Street Revival: Not a Miracle, Just Community Development." *Washington Post,* March 24, 2013.

Nicholls, Walter. "A New Chinatown." *Washington Post,* October 22, 2003.

Nicholls, Walter. "Washington's Little Ethiopia." *Washington Post*, May 18, 2005.

Nichols, David A. "'The Showpiece of Our Nation': Dwight D. Eisenhower and the Desegregation of the District of Columbia." *Washington History* (Fall/Winter 2004): 44–65.

Nora, Pierre. "Between Memory and History; Les Lieux de Mémoire." *Representations* 26 (Spring 1989).

Norton, Eleanor Holmes. "Home Rule Doesn't Come with an Asterisk." *Washington Post,* April 24, 2011.

Nuesse, C. Joseph. "Segregation and Desegregation at the Catholic University of America." *Washington History* 9:1 (Spring/Summer 1997): 54–70.

O'Connell, Jonathan. "Tysons Corner: The Building of an American City." *Washington Post,* September 24, 2011.

O'Connell, Jonathan. "Can City Life Be Exported to the Suburbs?" *Washington Post*, September 7, 2012.

O'Connell, Jonathan. "St. Elizabeth's Renovation as Security Campus Faces Resistance." *Washington Post,* March 30, 2012.

O'Flaherty, Brendan. *City Economics.* Cambridge, MA: Harvard University Press, 2005.

O'Hare, William P. and William H. Frey. "Booming, Suburban, and Black." *Demographics Magazine*, September 1992. http://www.frey-demographer.org/briefs/B-1992-2_BecomingSuburban.pdf.

O'Malley, Therese. "'A Public Museum of Trees': Mid-Nineteenth Century Plans for the Mall." In *The Mall in Washington, 1791–1991,* edited by Richard Longstreth. Washington, D.C.: National Gallery of Art, 1991.

"On the Capitol Steps." *Washington Post.* May 1, 1894.

Open Secrets. "Lobbying Database." 2011. Accessed August 23, 2011. http://www.opensecrets.org/lobby/index.php.

Oppel, Frank and Tony Meisel, editors. *Washington, D.C.: A Turn-of-the-Century Treasury.* Secaucus, NJ: Castle, 1987.

Orfield, Myron and Thomas Luce. 2012. "America's Racially Diverse Suburbs: Opportunities and Challenges." Institute of Metropolitan Opportunity, University of Minnesota Law School, July 20, http://www.law.umn.edu/uploads/5f/0b/5f0b8b86d389c4416a08bb29a3614ed2/Diverse_Suburbs_FINAL.pdf.

Ornstein, Norman J., Thomas E. Mann, and Michael J. Malbin. *Vital Statistics on Congress.* Washington, D.C.: Brookings Institution Press, 2008.

Pacheco, Josephine F. *The Pearl: A Failed Slave Escape on the Potomac.* Chapel Hill: University of North Carolina Press, 2005.

Pacifico, Michele F. "'Don't Buy Where You Can't Work': The New Negro Alliance of Washington." *Washington History* 6:1 (Spring/Summer 1994): 66–88.

Paris, Jenell Williams. "*Fides* Means Faith: A Catholic Neighborhood House in Lower Northwest Washington, D.C." *Washington History* 11:2 (Fall/Winter 1999/2000): 24–45.

Parr, Marilyn K. "Chronicle of a British Diplomat: The First Year in the 'Washington Wilderness.'" *Washington History* 12:1 (2000): 78–89.

Parsons, Talcott. *The Social System.* New York: The Free Press, 1951.

Paullin, Charles O. "Early British Diplomats in Washington." *Records of the Columbia Historical Society* 44/45 (1942): 241–262.

Pearson, Drew and Robert S. Allen. "Boiled Bosoms." In *Katharine Graham's Washington*, edited by Katharine Graham. New York: Vintage Books, 2002.

Pearson, Drew and Robert S. Allen. "The Society of the Nation's Capital." In *Katharine Graham's Washington*, edited by Katharine Graham. New York: Vintage Books, 2002.

Peets, Elbert. *On the Art of Designing Cities: Selected Essays of Elbert Peets.* Edited by Paul D. Spreiregen. Cambridge, MA: The M.I.T. Press, 1968.

Perlin, Ross. "Five Myths about . . . Interns." *Washington Post,* May 20, 2011.

Peterson, Jon A. "The Senate Park Commission Plan for Washington, D.C.: A New Vision for the Capital and the Nation." In *Designing the Nation's Capital: The 1901*

Plan for Washington, D.C., edited by Sue Kohler and Pamela Scott. Washington, D.C.: U.S. Commission of Fine Arts, 2006.

Peterson, Paul E. *City Limits.* Chicago: The University of Chicago Press, 1981.

Pew Research Center. "The New Washington Press Corps." 2009. Accessed September 9, 2011. http://www.journalism.org/analysis_report/numbers.

Phipps-Evans, Michelle. "What Makes a Strong DC Community—A Hillcrest Community Civic Association Oral History Project—Overview." *DC Digital Museum.* Accessed November 20, 2012. http://www.wdchumanities.org/dcdm/items/show/1518.

Pianin, Eric and Courtland Milloy. 1985. "Does the White Return to D.C. Mean 'The Plan' Is Coming True?" *Washington Post,* October 6.

Pianin, Eric and Saundra Torry. "Barry Seeks to Quell Anger of D.C. Officials." *Washington Post,* October 4, 1988.

Pickens, Buford. "Mr. Jefferson as Revolutionary Architect." *Journal of the Society of Architectural Historians* 34:4 (1975): 257–279.

Pierson, William H., Jr. *American Buildings and Their Architects, Vol. 1: The Colonial and Neoclassical Styles.* New York: Oxford University Press, 1970.

Places and Persons on Capitol Hill: Stories and Pictures of a Neighborhood. Washington, D.C.: Capitol Hill Southeast Citizens Association, 1960.

Popple, Phillip R. and Leslie Leighninger. *Social Work, Social Welfare, and American Society*, eighth edition. Upper Saddle River, NJ: Pearson, 2011.

"President Leaves Today." *Washington Post,* August 11, 1893.

President's Committee on Civil Rights. *To Secure These Rights: The Report of the President's Committee on Civil Rights*. 1948.

Press, Donald E. "South of the Avenue: From Murder Bay to the Federal Triangle." *Records of the Historical Society, Washington, D.C.* 51 (1984): 51–70.

PricewaterhouseCoopers. *UK Economic Outlook*. November 2009. Accessed November 5, 2012. http://www.ukmediacentre.pwc.com/imagelibrary/downloadMedia.ashx?MediaDetailsID=1567.

Provine, Dorothy. "The Economic Position of Free Blacks in the District of Columbia, 1800–1860." *The Journal of Negro History* 58:1 (1973): 61–72.

Pumar, Enrique S. editor. *Hispanic Migration and Urban Development: Studies from Washington DC.* Bingley, UK: Emerald Press, 2012.

"Race Issue up Again." *Washington Post,* March 7, 1914.

"Race Policy Problem." *Washington Post,* September 30, 1913.

Rae, Douglas W. *City: Urbanism and Its End.* New Haven: Yale University Press, 2005.

Reilly, Corinne. "Fairfax Already Feeling Chill of Cuts." *Washington Post,* October 24, 2012.

Reinalda, Bob. *Routledge History of International Organizations: From 1815 to the Present Day*. New York: Routledge, 2009.

Reps, John W. *The Making of Urban America: A History of City Planning in the United States.* Princeton, New Jersey: Princeton University Press, 1965.

Reps, John W. *Monumental Washington: The Planning and Development of the Capital Center.* Princeton, New Jersey: Princeton University Press, 1967.

Rich, Frank. "The De Facto Capital." *New York Times Magazine,* October 6, 2002.

Ricks, Mary Kay. "Escape on the *Pearl*." *Washington Post,* August 12, 1998.

Riley, Corinne and Victor Zapana. "Tysons Corner Is Unofficially Dropping the 'Corner' from Its Name." *Washington Post.* October 4, 2012.

Roberts, Roxanne and Amy Argetsinger. "At Nerd Prom, the Glitzy Set Grabs Control." *Washington Post,* May 2, 2011.

Roberts, Roxanne and Amy Argetsinger. "Lee's Ripe for Roasting." *Washington Post,* February 11, 2011.

Roberts, Roxanne and Amy Argetsinger. "Spotting the Quasi-(in)famous." *Washington Post,* February 14, 2011.

Roediger, David R. *The Wages of Whiteness: Race and the Making of the American Working Class.* New York: Verso, 1999.

Roig-Franzia, Manuel. "Researcher Links Slaves to Castle's Sandstone." *Washington Post,* December 13, 2012.

Rothstein, Edward. "A Mirror of Greatness, Blurred." *New York Times,* August 25, 2011.

Ruble, Blair A. *Washington's U Street: A Biography.* Washington, D.C.: Woodrow Wilson Center Press, 2010.

Ruth Ann Overbeck Capitol Hill History Project. Accessed September 20, 2012. http://capitolhillhistory.org/index.html.

Rybczynski, Witold. "'A Simple Space of Turf': Frederick Law Olmstead Jr.'s Idea for the Mall." In *The National Mall: Rethinking Washington's Monumental Core,* edited by Nathan Glazer and Cynthia R. Field. Baltimore: Johns Hopkins University Press, 2008.

Rybeck, Rick. "Using Value Capture to Finance Infrastructure and Encourage Compact Development." *Public Works Management and Policy* 8:4 (2004): 249–260.

Samuels, Robert. "Community Deluged by Sewage." *Washington Post,* August 26, 2012.

Sanchez, Rene. "Barry Comes Roaring Back in D.C." *Washington Post,* September 14, 1994.

Sandage, Scott. "A Marble House Divided: The Lincoln Memorial, the Civil Rights Movement, and the Politics of Memory, 1939–1963." *Journal of American History* 80:1 (June 1993): 135–167.

Savage, Kirk. *Monument Wars: Washington, D.C., The National Mall, and the Transformation of the Memorial Landscape.* Berkeley: University of California Press, 2009.

Scheiber, Walter A. "Washington's Regional Development." *Records of the Columbia Historical Society* 49 (1973/1974): 595–603.

Schrag, Phillip G. "The Future of District of Columbia Home Rule." *Catholic University Law Review* 39:2 (1990): 311–371.

Schrag, Zachary M. *The Great Society Subway: A History of the Washington Metro.* Baltimore: Johns Hopkins University Press, 2006.

Schulte, Brigid. "Wheaton Seeks Bridge across Cultures." *Washington Post,* February 15, 2011.

Scott, Pamela. *Temple of Liberty: Building the Capitol for a New Nation.* New York: Oxford University Press, 1995.

Scott, Pamela. "'This Vast Empire': The Iconography of the Mall, 1791–1848." In *The Mall in Washington, 1791–1991,* edited by Richard Longstreth. Washington, D.C.: National Gallery of Art, 1991.

Scully, Vincent. *American Architecture and Urbanism.* Revised edition. New York: Henry Holt and Company, 1988.

Scully, Vincent. *Architecture: The Natural and the Manmade.* New York: St. Martin's Press, 1991.

Segraves, Mark. "Diplomats in D.C. Owe Hundreds of Thousands in Parking Tickets." *WTOP*, September 21, 2011. Accessed October 2, 2012. http://www.wtop.com/41/2555713/Diplomats-in-DC-owe-hundreds-of-thousands-in-parking-tickets.

Selassie, Bereket H. "Washington's New African Immigrants." In *Urban Odyssey: A Multicultural History of Washington, D.C.,* edited by Francine Curro Cary. Washington, D.C.: Smithsonian Institution Press, 1996.

Sharoff, Robert. "At the World Bank, Architecture as Diplomacy." *New York Times,* March 9, 1997.

Shin, Annys. "Ethiopian Soccer Tournament Promoting Unity Leads to Division." *Washington Post,* July 5, 2012.

Shogan, Robert. *Harry Truman and the Struggle for Racial Justice.* Lawrence: University of Kansas Press, 2013.

Shugart, Matthew Soberg and John M. Carey. *Presidents and Assemblies: Constitutional Design and Electoral Dynamics.* New York: Cambridge University Press, 1992.

Simon, Nina. *The Participatory Museum.* Santa Cruz, CA: Museum 2.0, 2010.

Singer, Audrey. "Metropolitan Washington: A New Immigrant Gateway." In *Hispanic Migration and Urban Development: Studies from Washington DC,* edited by Enrique S. Pumar. Bingley, UK: Emerald Press, 2012.

Singer, Audrey. "The Rise of New Immigrant Gateways." *The Living Cities Census Series*, Washington, D.C.: Brookings Institution, February 2004.

Slack, Enid and Rupak Chattopadhyay. "Finance and Governance of Capital Cities in Federal Systems." Washington, D.C. Economic Partnership Forum, Washington, D.C., September 16, 2010. Accessed January 15, 2012. http://www.youtube.com/watch?v=XG-fY3HNuFU&feature=related.

Smith, Hedrick. *The Power Game: How Washington Works.* New York: Ballantine Books, 1988.

Smith, Kathryn Schneider, editor. *Washington at Home: An Illustrated History of Neighborhoods in the Nation's Capital*, second edition. Baltimore: The Johns Hopkins University Press, 2010.

Smith, Thomas G. *Showdown: JFK and the Integration of the Washington Redskins.* Boston: Beacon Press, 2011.

Solomon, Burt. *The Washington Century: Three Families and the Shaping of the Nation's Capital.* New York: HarperCollins, 2004.

Somers, Meredith. "Fight Goes on over Mall Visitors Center." *The Washington Times.* July 17, 2012.

Spirou, Costas. "Both Center and Periphery: Chicago's Metropolitan Expansion and the New Downtowns." In *The City, Revisited: Urban Theory from Chicago, Los*

Angeles, and New York, edited by Dennis R. Judd and Dick Simpson. Minneapolis: University of Minnesota Press, 2011.

Spitzer, Neil. "A Secret City." *Wilson Quarterly* 13:1 (New Year's, 1989): 102–115.

Starr, Kevin. *Golden Gate: The Life and Times of America's Greatest Bridge.* New York: Bloomsbury Press, 2010.

Stephenson, Richard W. *A Plan Whol[l]y New: Pierre Charles L'Enfant's Plan of the City of Washington.* Washington, D.C.: Library of Congress, 1993.

Stewart, James B. "Christian Statesmanship, Codes of Honor, and Congressional Violence: The Antislavery Travails and Triumphs of Joshua Giddings." In *In the Shadow of Freedom: The Politics of Slavery in the National Capital*, edited by Paul Finkelman and Donald R. Kennon. Athens: Ohio University Press, 2011.

Stewart, Nikita and Paul Schwartzman. "How Adrian Fenty Lost His Reelection Bid for D.C. Mayor." *Washington Post,* September 16, 2010.

Stillman, Damie. "From the Ancient Roman Republic to the New American One: Architecture for a New Nation." In *A Republic for the Ages: The United States Capitol and the Political Culture of the Early Republic*, edited by Donald R. Kennon. Charlottesville, VA: University Press of Virginia, 1999.

Stolberg, Sheryl Gay. "Daschle, Democratic Senate Leader, Is Beaten." *New York Times,* November 3, 2004.

Streatfield, David C. "The Olmsteads and the Landscape of the Mall." In *The Mall in Washington, 1791–1991,* edited by Richard Longstreth. Washington, D.C.: National Gallery of Art, 1991.

Sugrue, Thomas J. *Origins of the Urban Crisis: Race and Inequality in Postwar Detroit.* Princeton, NJ: Princeton University Press, 1996.

Sunlight Foundation. "Keeping Congress Competent: Staff Pay, Turnover, and What It Means for Democracy." 2010. Accessed November 21, 2011. http://assets.sunlightfoundation.com.s3.amazonaws.com/policy/papers/Staff%20Pay%20Blogpost.pdf.

Sweeney, Michael S. "'The Desire for the Sensational': Coxey's Army and the Argus-Eyed Demons of Hell." *Journalism History*, 23:3 (1997): 114–125.

Tajfel, Henri and John C. Turner. "The Social Identity Theory of Inter-group Behavior." In *Psychology of Intergroup Relations*, edited by Stephen Worchel and William G. Austin. Chicago: Nelson-Hall Publishers, 1986.

Tate, Katherine. *From Protest to Politics: The New Black Voters in American Politics.* New York: Russell Sage Foundation, 1993.

Tatian, Peter A., G. Thomas Kingsley, Margery Austin Turner, Jennifer Comey, and Randy Rosso. *State of Washington, D.C.'s Neighborhoods.* Washington, D.C.: The Urban Institute, 2008. Accessed December 9, 2013. http://www.urban.org/uploadedpdf/411881_stateofwashington.pdf.

Tavernor, Robert. *Palladio and Palladianism.* London: Thames & Hudson Ltd., 1991.

Taylor, Kate. "National Latino Museum Plan Faces Fight." *New York Times*. April 20, 2011.

Taylor, Steven. "Political Culture and African Americans' Forgiveness of Elected Officials." *Polity* 37:4 (October 2005): 491–510.

Teaford, Jon C. *The Metropolitan Revolution.* New York: Columbia University Press, 2006.

Terrell, Mary Church. "What It Means to Be Colored in Capital of U.S." 1906. Accessed November 20, 2012. http://www.americanrhetoric.com/speeches/marychurchterellcolored.htm.

Teute, Fredrika J. "Roman Matron on the Banks of the Tiber River." In *A Republic for the Ages: The United States Capitol and the Political Culture of the Early Republic*, edited by Donald R. Kennon. Charlottesville, VA: University Press of Virginia, 1999.

Thadani, Dhiru A. *The Language of Towns and Cities: A Visual Dictionary.* New York: Rizzoli, 2010.

The American Heritage Dictionary of the English Language, third edition. New York: Houghton Mifflin, 1992.

The District of Columbia. "Advisory Neighborhood Commissions." Accessed November 20, 2012. http://anc.dc.gov/.

The Museum of Broadcast Communications. "The Civil Rights Movement and Television." Accessed December 7, 2008. http://www.museum.tv/eotvsection.php?entrycode=civilrights.

The Urban Institute. "Neighborhood Profiles." 2012. Accessed January 9, 2013. http://www.neighborhoodinfodc.org/ wards/wards.html.

Thomas, Christopher A. *The Lincoln Memorial and American Life.* Princeton, NJ: Princeton University Press, 2002.

Thompson, J. Phillip, III. *Double Trouble: Black Mayors, Black Communities, and the Call for a Deep Democracy.* New York: Oxford University Press, 2006.

Thompson, Krissah. "Michelle Obama's Washington." *Washington Post,* September 26, 2013.

Tocqueville, Alexis de. *Democracy in America,* J. P. Mayer, ed. New York: Harper & Row, 1969 [1831].

Toeplitz, Shira. "Ex-Rep. Halvorson May Challenge Jackson." *Roll Call,* September 6, 2011.

Transportation Plan; National Capital Region. The Mass Transportation Survey Report, 1959. Washington, D.C.: National Capital Planning Commission, 1959.

Trescott, Jacqueline. "Ant-Covered Jesus Video Removed from Smithsonian after Catholic League Complains." *Washington Post,* December 1, 2010.

Trescott, Jacqueline. "City Museum to Close Its Galleries." *Washington Post.* October 9, 2004.

Trollope, Frances. "The Domestic Manners of the Americans." In *Literary Washington, D.C.,* edited by Patrick Allen. San Antonio: Trinity University Press, 2012.

Tsui, Bonnie. "The End of Chinatown." *The Atlantic,* December 2011.

Turner, John C., with Michael A. Hogg, Penelope J. Oakes, Stephen D. Reicher, and Margaret S. Wetherell. *Rediscovering the Social Group: A Self-Categorization Theory.* New York: Basil Blackwell, 1987.

Tuskegee University Archives Online Repository. 2010. *Lyching, Whites & Negroes, 1882–1968.* [Data file]. Retrieved from http://192.203.127.197/archive/bitstream/handle/123456789/463/Lyching%201882%201968.pdf?sequence=1

Twomey, Steve. "Blacks Make Barry the People's Chance." *Washington Post,* September 15, 1994.

Tydings, Joseph D. "Home Rule for the District of Columbia: The Case for Political Justice." *American University Law Review* 16 (1967): 271–277.

U.S. Census Bureau. *Patterns of Metropolitan and Micropolitan Population Change: 2000 to 2010.* September 2012. Accessed December 9, 2013. http://www.census.gov/prod/cen2010/reports/c2010sr-01.pdf.

U.S. House of Representatives. Committee on the District of Columbia. *Building Height Limitations.* Serial No. S-5. Washington, D.C.: Government Printing Office, 1976.

U.S. House of Representatives, Chief Administrative Office. *2010 House Compensation Study.* Washington, D.C.: ICF International, 2010.

U.S. Office of Personnel Management. "Table 1—Executive Branch (non-Postal) Employment by Gender, Race/National Origin, Disability Status, Veterans Status, Disabled Veterans." September 2006. Accessed September 9, 2011. http://www.opm.gov/feddata/demograp/table1-1.pdf.

U.S. Office of Personnel Management. "Table 2—Comparison of Total Civilian Employment of the Federal Government by Branch, Agency, and Area as of August 2009 and September 2009." September 2009. Accessed August 22, 2011. http://www.opm.gov/feddata/html/2009/September/table2.asp.

U.S. Office of Personnel Management. "Table 1—Race/National Origin Distribution of Civilian Employment, Executive Branch Agencies, World Wide." September 2006. Accessed September 9, 2011. http://www.opm.gov/feddata/demograp/table1mw.pdf.

"Upholds Race Purity." *Washington Post,* February 11, 1913.

Van de Water, Frederic. "The Society of the Nation's Capital." In *Katharine Graham's Washington,* edited by Katharine Graham. New York: Vintage Books, 2002.

Van Dyne, Larry. "Foreign Affairs: DC's Best Embassies." *Washingtonian Magazine,* February 1, 2008.

Veblen, Thorstein. *The Theory of the Leisure Class.* New York: The MacMillan Company, 1899.

Vitruvius. *On Architecture.* Translated by Richard Schofield. New York: Penguin Books, 2009.

Warner, Sam Bass, Jr. *The Urban Wilderness: A History of the American City.* Berkeley: University of California Press, 1995 [1972].

"Washington DC's 2011 Visitor Statistics." *Destination DC,* 2011. Accessed February 11, 2013. http://planning.washington.org/images/pdfs/2011_VisitorStatistics2.pdf.

"Washington Post 200." *Washington Post,* 2010. Accessed December 9, 2013. http://www.washingtonpost.com/wp-srv/special/business/post200-2009/post-200-graphic.html.

"Washington Wire." *Wall Street Journal,* May 16, 1975.

Wax, Emily. "Black Middle Class Is Redefining Anacostia." *Washington Post,* July 29, 2011.

Wax, Emily. "Ethiopian Yellow Pages: Life, by the Book." *Washington Post,* June 8, 2011.

Wax, Emily. "Outpost Betting a Nation." Washington Post, December 27, 2011.

Wax, Emily. "Washington Can Be a Frontline for International Combatants." *Washington Post,* January 11, 2012.

Weeks, Linton. "Maya Lin's 'Clear Vision.'" *Washington Post,* October 20, 1995.

Weiss, Eric M. "Chiseling away at Chinatown." *Washington Post,* February 14, 2005.

Weiss, Nancy J. *Farewell to the Party of Lincoln: Black Politics in the Age of FDR.* Princeton: Princeton University Press, 1983.

Westley, Brian. 2005. "Washington: Nation's Largest Ethiopian Community Carves Niche." *Associated Press,* October 17.

"What's the Story on Diplomatic Immunity?" *The Straight Dope,* November 1, 2005.

Whiffen, Marcus and Frederick Koeper. *American Architecture, Volume 1: 1607–1860.* Cambridge, MA: The M.I.T. Press, 1981.

White House. *Annual Report to Congress on White House Staff.* 2011. Accessed August 23, 2011. http://www.whitehouse.gov/briefing-room/disclosures/annual-records/2011.

Whyte, James H. *The Uncivil War: Washington during the Reconstruction, 1865–1878.* New York: Twayne Publishers, 1958.

Widdicombe, Gerry. "The Fall and Rise of Downtown D.C." *Urbanist,* January 2010. http://www.spur.org/publications/library/article/fall_and_rise_downtown_dc.

Wilkerson, Isabel. *The Warmth of Other Suns: The Epic Story of America's Great Migration.* New York: Random House, 2010.

Williams, Brett. *Upscaling Downtown: Stalled Gentrification in Washington, D.C.* Ithaca: Cornell University Press, 1988.

Williams, Juan. *Eyes on the Prize: America's Civil Rights Years, 1954–1965.* New York: Penguin Books, 1988.

Williams, Paul A. *Greater U Street.* Charleston, SC: Arcadia Publishing, 2001.

Williamson, Mary Lou, ed. *Greenbelt: History of a New Town, 1937–1987.* Norfolk, VA: Dunning, 1987.

Wilson, Jill H. and Audrey Singer. *State of Metropolitan America.* Washington, D.C.: The Brookings Institution, October 2011.

Wilson, Richard Guy. "High Noon on the Mall: Modernism versus Traditionalism, 1910–1970." In *The Mall in Washington, 1791–1991,* edited by Richard Longstreth. Washington, D.C.: National Gallery of Art, 1991.

Wilson, William H. *The City Beautiful Movement.* Baltimore, MD: Johns Hopkins University Press, 1994.

Wolman, Hal, Jan Chadwick, Ana Karruz, Julie Friendman, and Garry Young. "Capital Cities and Their National Governments: Washington, D.C. in Comparative Perspective." GWIPP Working Paper 30, June 11, 2007. Accessed January 4, 2012. http://www.gwu.edu/~gwipp/papers/wp030.pdf.

Yarwood, Doreen. *Encyclopedia of Architecture.* New York: Facts on File Publications, 1986.

Young, James Sterling. *The Washington Community, 1800–1828.* New York: Columbia University Press, 1966.

Young, Tara. "Diplomat Flees U.S. to Avoid Sex Charges." *Washington Post,* March 5, 2005.

Yu, Meeja and Unyong Kim. "'We Came Here with Dreams': Koreans in the Nation's Capital." In *Urban Odyssey: A Multicultural History of Washington, D.C.,* edited by Francine Curro Cary. Washington, D.C.: Smithsonian Institution Press, 1996.

Zelinsky, Wilbur and Barrett A. Lee. "Heterolocalism: An Alternative Model of the Sociospatial Behavior of Immigrant Ethnic Communities." *International Journal of Population Geography* 4:4 (December 1998): 281–298.

Zak, Dan. "Party Politics on Tap." *Washington Post,* June 6, 2011.

Index

Printed and bound in the United States of America